DEVELOPMENT AND PLANNING OF CHINESE CULTURAL AND COMMERCIAL STREETS
——MODERN RIVERSIDE SCENE AT QINGMING FESTIVAL

中国文化商业古街
开发与规划资料集
——现代清明上河图

金盘地产传媒有限公司 策划

唐艺设计资讯集团有限公司 编

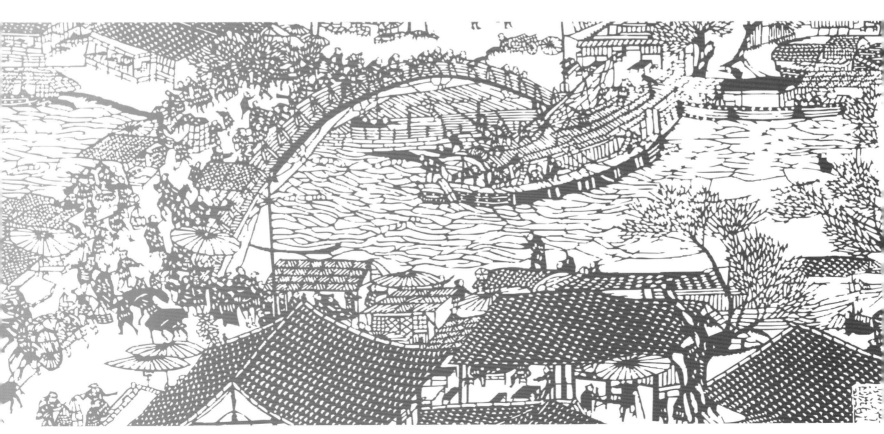

海峡出版发行集团　福建科学技术出版社
THE STRAITS PUBLISHING & DISTRIBUTING GROUP　FUJIAN SCIENCE & TECHNOLOGY PUBLISHING HOUSE

古城镇扩建的商业街区

Commercial Streets Extended from Ancient Towns

022-129

下

山西平遥南大街
Pingyao South Street, Shanxi

024

山西朔州老街
Shuozhou Old Street, Shanxi

036

苏州同里
Tongli, Suzhou

056

丽江大研古城
Dayan Old Town, Lijiang

108

painted rafters. Since local residents pay much attention to façade, most buildings are designed with single or double eaves, wooden structure roofs and overhanging brackets. Each household features courtyards as well as orchid flowers.

— Nanzhao Ancient Street, Dali

There are rows upon rows of Naxi style folk houses in the old town. The exquisite arch bridges in front of each building span across crystal streams running over the town, while willow branches beside water dance gently in the breeze. Black stone lanes zigzag throughout the town and connect with each other. As a result, unique water town scenery on plateau is created.

— Dayan Old Town, Lijiang

The general design of Confucius Temple Qinhuai Scenic Area stresses traditional Chinese architectural elements and covers or forbids modern stuffs such as air conditioner, and A-alloy doors and windows.

— Confucius Temple, Nanjing

Window frames are all painted red and contrasted with stone road on the street to form a distinct color effect. Buildings mostly apply wooden structure. Their lintels are carved with lane names such as "Chang'an Li", "Jirui Li" and "De'an Li". The houses are in ancient residential style with a courtyard in the middle.

— Xijin Ferry Street, Zhenjiang

Corridors and quays are equipped along the river channel revealing that people depended on river channel in the past. Buildings away from the river are organized by courtyards of different sizes with changeable space layout.

— Cangqiao Zhijie, Shaoxing

Buildings here generally adopt local stones and red bricks. Their materials are selected and their and gardens to construct a unique appearance of Tongli Old Town. The layout of bridges, streams and dwellings wins a praise of "oriental small Venice".

— Tongli, Suzhou

The bridges and streets are connected, and the buildings are along the rivers. More than 60% of houses are Ming and Qing dynasties residential buildings, and there are nearly 100 classic houses and more than 60 classic brick gatehouses, which are all of primitive beauty. There are 14 distinctive bridges, and together with the water and people, they create a wonderful watery landscape. Four main rivers intersect with each other, and form a perfect landscape.

— Zhouzhuang, Suzhou

The four river-side streets are paralleled, and water and land are adjacent. Every corner embodies a harmonious overall beauty of humanistic and natural environment. The houses are closely linked to the river, together with other ancient buildings making a unique charm of bridges, water and old houses.

— Wuzhen, Tongxiang

The Tulouba, Sanheping and Mini Shanghai are courtyard complexes mingling Sichuan, Fujian, Shanghai, Shanxi and Anhui architectural styles. Every household is independent from each other, yet they all show luxurious decoration. Winding wide streets and narrow lanes outline the leisure life of the town and hide prosperity inside.

— Boke Town, Chengdu

Most folk houses in the district have retained the original features of Ming and Qing dynasties. An ancient street connects four old towers and forms harmonious landscapes. All buildings are simple and elegant, with carved beams and

044
上海七宝老街
Qibao Old Street, Shanghai

066
苏州周庄
Zhouzhuang, Suzhou

084
桐乡乌镇
Wuzhen, Tongxiang

098
大理南诏古街
Nanzhao Ancient Street, Dali

122
成都洛带博客小镇
Boke Town, Chengdu

现代仿建的商业街区

Modern Imitated Commercial
Streets of Ancient Style

022-262

中

入。同时对现代建筑符号，比如空调外挂机、铝合金门窗等予以遮挡或取缔。修缮的建筑按照『修旧如旧』的原则，采用原有建筑材料和色彩。

——南京夫子庙

飞檐雕花的窗栏一律油漆成朱红色，给人以『飞阁流丹』之感，形成了『前店后寝』或『下面为店铺，楼上为住家』的建筑形式，和街面的青石板路相互衬托，形成鲜明的色彩效果。建筑大多采用木结构，按里弄聚居。券门的门楣上镌刻着『长安里』『吉瑞里』『德安里』等字样。四合院中间为天井，反映出江南特有的『四水归堂』古民居建筑风格。

——镇江西津渡古街

临水建筑设沿廊、埠头，以传统的绍兴木格子窗为主，沿河的阳台则恢复成披檐，反映了过去人民生活对水道的依赖。不临水的建筑由多户住宅形成连片建筑，以大小不等的院落（天井）组织平面，并由狭窄廊道相连，这样的街坊布局空间变化较多。

——绍兴仓桥直街

shapes are unique. The facades of buildings are decorated with Romanesque columns and slope roofs of totally different shapes. Some of the buildings imitate ancient palatial architecture while blending Western European architectural technique to create a compromise style of unique appearance.

— Gulangyu Island, Xiamen

建筑一般选用当地的石材或红砖，用料考究、造型别致，且多地，辅以奇异别致的琉璃瓶花格。建筑的各个立面常精雕细刻罗马采用圆拱回廊、清水红砖、红瓦坡屋面，并用柚木楼板、花砖铺式大型圆柱和结构造型迥然不同的多坡屋顶。有的建筑仿照古代宫殿式建筑并融合西欧建筑造型手法，形成外形独特、屋檐线条奇异的折中式建筑风格。

——厦门鼓浪屿

上海新天地
Xintiandi, Shanghai
074

北京前门大街
Qianmen Street, Beijing
100

黄山置地国际中心
Huangshan Land International Center
200

扬州东关街
Dongguan Street, Yangzhou
258

宁波南塘老街
Nantang Street, Ningbo
272

福州三坊七巷
Sanfang Qixiang, Fuzhou
288

巧妙而自然地把水、路、桥、民居、园林等融为一体，构成了古镇同里特有的水乡风貌。镇内家家临水，户户通舟；明清民居鳞次栉比；宋、元、明、清各个时期的桥保存完好。以小桥、流水、人家的格局赢得『东方小威尼斯』的美誉。

——苏州同里

桥街相连，依河筑屋，全镇60%以上的民居仍为明清建筑，有近百座古典宅院和60多个砖雕门楼。同时，还保存了14座各具特色的古桥，共同勾勒了一幅美妙的『小桥、流水、人家』的水乡风景画。南北市河、后港河、油车漾河、中市河形成『井』字形，傍水筑屋，依河成街，深宅大院，重脊高檐，河埠廊坊，过街骑楼，穿竹石栏，临河水阁，一派古朴幽静的景象。

——苏州周庄

古风犹存的东、西、南、北四条老街呈十字交叉，构成双棋盘式河街平行、水陆相邻的古镇格局。民居宅屋傍河而筑，街道两旁保存有大量明清建筑，辅以河上石桥，体现了小桥、流水、古宅的江南古镇风韵。

——桐乡乌镇

土楼坝、三和坪、小上海均为院落式商业集群，融川、闽、海、晋、徽五派建筑风格于一体。每栋建筑均为藏品级定制，以古都名宅为用材标准，由名匠古法精工打造，一砖一木，皆精湛厚重。一院一字号，户户皆流金。宽街窄巷的迂回曲折，勾勒出悠然的小镇生活，藏繁华于内，立名门家业。

——成都洛带博客小镇

民居大多保留了明清建筑风貌，多为土木结构。青瓦坡顶、『三坊一照壁』『四合五天井』，古街串联四座古楼，头尾衔接，相得益彰。建筑古朴典雅，雕梁画栋。民居讲究门面，大门多为单檐或重檐，木架瓦顶，出阁架斗，户户有庭院，家家栽兰花，是『满城春兰风亦香』的真实写照。

——大理南诏古街

古城纳西风格浓郁的民居鳞次栉比，家家门前精巧的拱桥横跨穿城过巷的清澈溪水，排排垂柳在轻风中摇曳，青石板铺就的小巷蜿蜒曲折、纵横交错，形成了『家家泉水，户户垂杨』与『小桥、流水、人家』的独特高原水乡风貌。

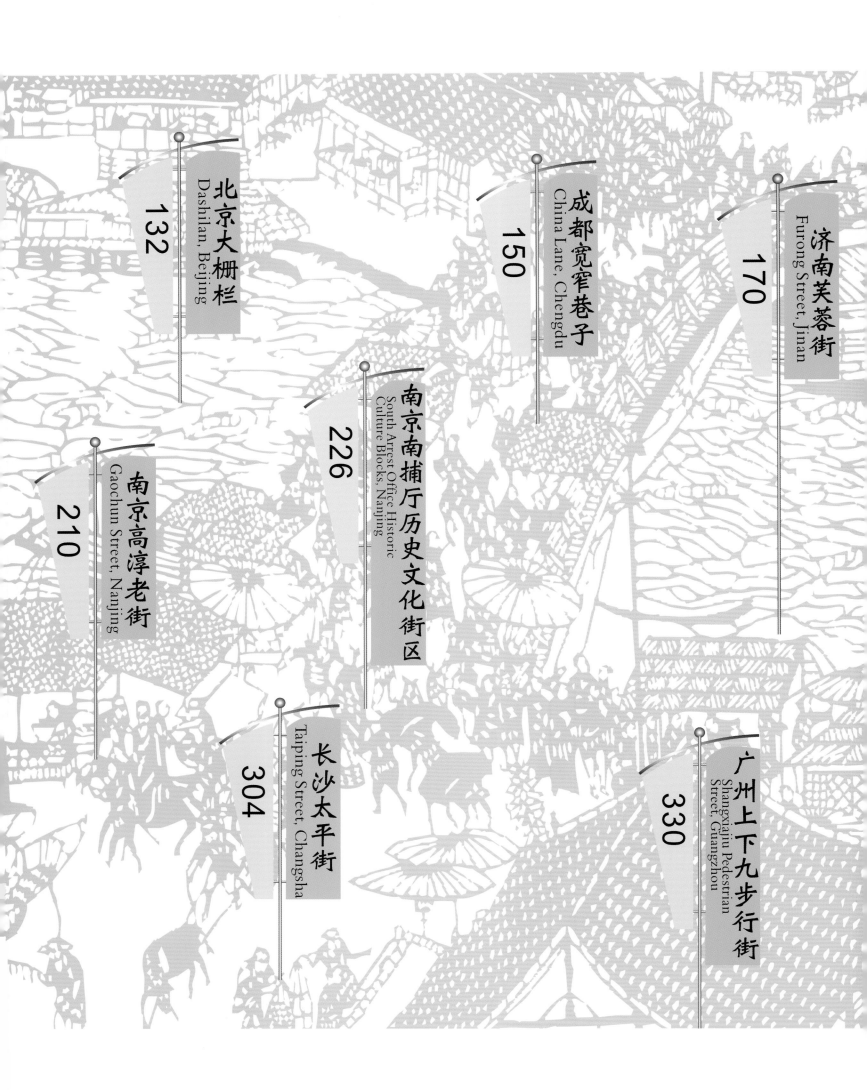

北京大栅栏
Dashilan, Beijing
132

成都宽窄巷子
China Lane, Chengdu
150

济南芙蓉街
Furong Street, Jinan
170

南京高淳老街
Gaochun Street, Nanjing
210

南京南捕厅历史文化街区
South Arrest Office Historic Culture Blocks, Nanjing
226

长沙太平街
Taiping Street, Changsha
304

广州上下九步行街
Shangxiajiu Pedestrian Street, Guangzhou
330

古住宅和古街道改建的商业街区

Commercial Streets Reconstructed from Ancient Residential Areas and Ancient Streets

022-339

上

佛山岭南天地
Lingnan Tiandi, Foshan

026

武汉天地
Wuhan Tiandi

052

黄山屯溪老街
Tunxi Street, Huangshan

182

苏州观前街
Guanqian Street, Suzhou

248

现代清明上河图

——项目经典片段摘录

整条大街由市楼分隔为南、北街，商铺建筑及辅院相连的基本格调一致，而又绝不雷同。高大宽绰的清代铺面，建筑风格考究，雕刻与绘画手法细腻精致，历经百年而古风犹存。

——山西平遥南大街

七宝老街拥有千年历史，整修后的老街两旁多是两层砖木混合结构的仿明清式建筑，粉墙黛瓦，配有深红色的木质门窗。宋代遗存的狭长的老街深巷上铺着青砖石板，蒲汇河与河上的三座小桥构成了江南小桥流水的景色。

——上海七宝老街

古城镇扩建的商业街区
Commercial Streets Extended
from Ancient Towns

022-129

CONTENTS 目录

Preface

序

Regional districts owning certain cultural deposits or traditional features are injected with cultural, leisurely and creative elements, so as to be renovated, extended or reconstructed as "international, cultural and fashionable" leisure commercial streets.

In recent years, after commercial real estates' rapid expansion, cultural commercial streets, with specific historical and cultural background, specific consumers and rich business activities, and combined with traditional culture, history and modern civilization, are popular in commercial real estate market. Their new development mode receives great popularity and is identified as a more intimate and sustainable mode.

With great popularity, domestic cultural commercial streets emerge in large numbers. On one hand, domestic commercial real estate magnates such as Wanda Group and Shui On Group all spare no expense to develop cultural commercial districts. On the other hand, unique historical and cultural deposits differentiate the cultural commercial districts from traditional commercial streets and form their own developing features and operation modes. In such a situation we edit this book themed with commercial street market positioning, planning and design, and operation, which includes many successful cultural commercial district projects on market, so as to provide all developers and designers interested in cultural commercial streets development with good examples for reference.

This publication includes three volumes with 44 representative cases collected from 30 cities of 18 provinces and municipalities in China. All cases are of high quality and wide scope, with high-definition pictures provided by our professional photographers on the spot.

According to development characteristics, the book is divided into 4 categories, i.e. commercial streets reconstructed from ancient residential areas and ancient streets, commercial streets extended from ancient towns, modern imitated commercial streets of ancient style, and commercial streets reconstructed on the basis of historic preservations, to present the development status and planning design features of domestic cultural commercial streets from all aspects.

There are 7 sections to introduce each case. From street background and market positioning, planning, design features, commercial activities and operation, to brand shops and cultural facilities, all detail descriptions are arranged for holistic presentation of the development and operation process of each case to provide general developers and designers with new and practical references.

  文化商业街区，是指在有一定文化底蕴或传统特色的区域地段，注入"文化、休闲、创意"元素，改造、扩建或仿建出具有"国际性、文化性、时尚性"的休闲娱乐商业街区。

  近年来，商业地产在经历同质化放量增长阶段后，文化商业街区这种以特定的历史文化为背景、以特定的消费人群为导向、以丰富的时尚业态为支撑、融合传统历史文化与现代文明的商业街区，受到商业地产市场的青睐。这种新型的商业地产开发模式受到业界的追捧，被业界认定为一种更具亲和力和可持续发展的经典模式。

  在商业地产市场和业界的双重青睐下，国内文化商业街区开发日趋火热。一方面，国内商业地产大鳄如万达集团、瑞安集团、1912集团等纷纷斥巨资打造文化商业街区项目，商业地产市场上涌现出了一批成功的文化商业街区开发案例，如佛山岭南新天地、宽窄巷子、楚河汉街、上海新天地、南京1912等；另一方面，由于文化商业街区独特的历史文化底蕴，使文化商业街区的开发理念与开发模式有别于传统的商业街区，形成自己的开发特点、运营模式。在此种局势下，我们策划了此套以文化商业街区的定位、规划、设计、运营为主题的图书，收录市场上成功的文化商业街区开发项目，旨在为有志于文化商业街区开发的开发商与设计师提供参考和借鉴。

  《中国文化商业古街开发与规划资料集》分上、中、下三册，共收录北京、天津、上海、重庆、山东、江苏、浙江、福建、广东、湖北、湖南、四川等18个省市30个城市的44个代表性项目，项目质量高、辐射范围广、数量大，项目图片全部由专业摄影师实地拍摄。

  本书中，我们将项目按开发特点划分为四大类，分别是古住宅和古街道改建的商业街区、古城镇扩建的商业街区、现代仿建的商业街区和以文物保护为重点改建的商业街区，全方位展现我国文化商业街区的开发现状和规划设计特色。

  在每个街区的编排中，我们从街区背景与定位、街区规划、设计特色、商业业态、市场运营、品牌商铺、文化设施七个方面详细介绍，力求全面呈现项目开发运营全过程，为广大开发商和设计师提供鲜活而极具实用性的文化商业街区参考案例。

以文物保护为重点改建的商业街区

Commercial Streets Reconstructed - on the Basis of Historic Preservations

130-206

132 南京夫子庙 Confucius Temple, Nanjing

154 镇江西津渡古街 Xijin Ferry Street, Zhenjiang

174 绍兴仓桥直街 Cangqiao Zhijie, Shaoxing

186 厦门鼓浪屿 Gulangyu Island, Xiamen

The basic styles of shop buildings connecting with courtyard are the same, but not identical. The great shops are delicately designed with exquisite carving and painting techniques, full of beauty under a long history.

— Pingyao South Street, Shanxi

The Qibao Old Street has over one thousand years of history. After renovation, it still shows two-storey Ming and Qing dynasties buildings built by bricks and timber, with plaster walls, black roof tiles and dark red wooden doors and windows. Narrow and deep ancient streets and lanes that were paved with black bricks are remains of Song Dynasty. Three bridges over the Puhui River and the river itself form unique Jiangnan water town scenery with water running beneath the bridges.

— Qibao Old Street, Shanghai

The design smartly integrates water, roads, bridges, residences

So many cities at home and abroad are famous hundred-year old towns. However, under the background of rapid economic development, these traditional old towns are faced with decaying danger. Therefore, to develop commercial streets inside old cities and towns of historic value has become a significant point in interior adjustment of city space, and its positioning, development, planning, design and operation are closely related to tourism and culture to a great extent.

Concept taking tourism development as principal thing

With the shift of economic focus, the earliest business and trade zone and blocks featuring historic buildings and cultural landscape have a great attraction toward tourists. Therefore, the focal point of commercial street development is transferred to sustainable tourism real estate. In the development, attention is paid to diversity and individuation of tourist products. For example, the Six Old Towns of Jiangnan, though share the sight of "small bridge, streams, houses", they have their own features. Zhouzhuang's streets and houses layout is a typical "river-street mode" Jiangnan market town; Tongli's mansions or courtyards present a quiet environment; Wuzhen's typical "house right beside river" shows the appearance described by Mao Dun in *The Shop of the Lin Family*.

Giving prominence to folk culture connotation of old towns

Featured culture of old towns is positioned as theme and general style of tourist products and culture connotation is taken as core competitiveness of tourist products. On the basis of reasonable protection, tourist products are integrated to highlight cultural features of old towns. Emphasis is laid on intangible culture relic protection to further improve culture connotation of old towns.

Meanwhile, "folk features" have become the primary factor that attracts visitors and a focus of tourism development. In the process of development, spatial form, art style and national tradition are emphasized, and uniqueness of folk culture is explored to transform folk custom into tourism products and manifest unique charm of the street.

Protection for material space of old towns

The most important condition for old cities and towns to exist is a certain number of historic buildings which have unique tradition material spaces and form strong spatial level and order. In the planning design, distinctive and unique landmark facilities are often set at entrances or important nodes to bring a strong impression as well as organize the space.

Away from urban noise and confusion, quiet, distant, decorous or antique, old towns have become beautiful tourist scenery line in rapid economic development. Plain old building, unique delicacies, various folk customs, profound culture and invariable living style make these commercial streets "living fossils" of ancient oriental civilization.

国内外许多城市都是著名的百年老镇。然而，随着经济的迅速发展、城市空间结构的重组、经济重心的转移，这些传统老镇非但不能恢复往日的繁华，反而面临衰败的危险。为此，在具有历史价值的古城镇中开发新的商业街，成为城市空间内部调整的一个举足轻重的手段，从而导致其定位、开发、规划、设计、运营在很大程度上与旅游、文化密切相关。

以旅游业开发为主的理念

随着经济重心的转移，古城镇最早的商贸区和以历史建筑、文化景观为特色的街区，其独特的建筑和文化对游客具有很大的吸引力。因此，商业街开发的重心转移到可持续发展的旅游地产上，形成了以旅游业开发为主的模式。在开发中，注重旅游产品的差异化和个性化。如同处江南的六大古镇，虽共同享有"小桥、流水、人家"的水乡意境，但仍各具特色。周庄的前街后河、前店后房，属于典型的"河街式"贸易集镇；同里的富家大园或者雅致小院，体现出宁静的水乡居住环境；甪直的唐代大寺庙，是以庙兴镇；南浔丝绸兴旺、巨贾辈出，是工商业托起的古镇；乌镇典型的"人家尽枕河"，体现出茅盾笔下《林家铺子》的风情；西塘以酒兴市，体现的是买酒饮酒的商业文化。

重点突出古镇的民俗文化内涵

将古镇特有的文化定位为旅游产品的主题及整体风格，并将文化内涵作为旅游产品的核心竞争力。强调在合理保护的基础上，将旅游产品进行整合，以突显古镇的文化特色。注重非物质文化遗产的保护，为当地各类民俗表演、戏曲表演、手工艺术等提供发展平台，并开展相应的旅游节庆文化活动，以进一步提升其文化内涵。

同时，"民俗特色"作为古镇吸引游人的首要因素，成为旅游开发的重点。在开发过程中，注重其自身的空间形式、艺术风格及民族传统的传承，并深刻挖掘其自身民族文化的不可复制性，将独有的民俗风情转化为旅游产品，从而彰显街区与众不同的魅力。

对古镇物质空间的保护

古城镇拥有最重要的条件就是：镇上集中了一定数量的历史建筑，具有城市独特的传统物质空间环境，形成强烈的空间层次和秩序。在规划设计中，常于入口处或重要的节点设置鲜明独特的标志性设施，一方面起到组织空间序列的作用，另一方面给人以强烈的印象，并产生亲切感。

远离都市的喧嚣和纷扰，或宁静、或悠远、或厚重、或古朴的老镇成为经济高速发展中一道亮丽的旅游风景线。质朴的古镇建筑、独有的美食佳肴、丰富的民俗节日、深厚的人文积淀、亘古不变的生活方式使商业街成为东方古老文明的"活化石"。

Pingyao South Street, Shanxi
山西平遥南大街

Street Background & Market Positioning

街区背景与定位

History 历史承袭

Pingyao South Street is one of the most prosperous streets in Pingyao Ancient City with a length of 690 meters. As early as the mid-Ming, Pingyao South Street has a commercial scale with nearly 80 shops. After going through the commercial heyday of Qing Dynasty, Pingyao South Street slumped due to the war and poverty. The structure of some shops changed after business transformation in the 1950s, but the traditional structure of most shops remained. With the development of individual and private economy after the 1980s, individual shops flourished and some dwellings were back to their original shops. In the 1990s, shops in the main street were repaired according to the ancient city protection planning requirements in accordance with the principle of "whole new as the old", to keep the original structure, raw materials, and crafts. In 1995, the Pingyao county government restored the Ming and Qing style of the Pingyao South Street's shops as one of ten projects in the county. Some features, including black bricks and grey tiles, wood-panel doors, reoiled paintings, old plaques, lanterns wore highlighted, which business curtains were highlighted, which cost more than 500,000 Yuan. In June 2009, the world heritage, Pingyao South Street of Pingyao Ancient City, was selected as the first "Chinese Historic and Cultural Street".

山西平遥南大街是平遥古城中最繁荣的一条街道，街道全长690米。早在明代中期，南大街的商业就已成规模，拥有将近80家商铺；到清代商业鼎盛时期，南大街商铺种类涵盖了各行业，南大街盛极一时；至"民国"时期，由于战乱和贫困的影响，城内市场萧条，南大街商业萎缩；新中国成立后，经过20世纪50年代的商业改造，部分商铺公私合营，多数店铺变成国营商店，有的也改为民居，一些店铺结构发生变化，然而大多店铺的传统建筑结构基本保持原状；20世纪80年代，随着个体和私营经济的发展，个体店铺蓬勃发展，原来改为民居的店铺恢复为商用铺面；到20世纪90年代，根据古城保护规划的要求，各主要街道的店铺按照"整新如旧"的原则进行修葺，保持原格局、原材料、原工艺；1995年，平遥县县政府将"恢复南大街店铺的明清风貌"列入全县十大工程之一，铺面维修突出青砖青瓦，木板门特色，并重新油饰彩画，制作、悬挂老字号牌匾、宫灯、商幌等，共耗资50余万元；2009年6月，世界文化遗产"平遥古城南大街"入选首批"中国历史文化名街"。

Location 区位特征

Pingyao South Street is located in Pingyao Ancient City, a famous cultural city with a long history of more than 2,700 years. As the central axis of the ancient city, extending from the north of the joint of East Street and West Street to South Gate, Pingyao South Street has lots of old and traditional boutiques on both sides, bustling and traditional; it has good tourism resources, also its cultural relics and historic sites, well-preserved historic buildings, relatively complete traditional structure and unique historic character, making it a and certain scale. Its economic culture is vital, and continues the original historic culture and social life and maintains the original social function and vitality.

平遥南大街位于拥有2 700多年历史的文化名城——平遥古城内，是平遥古城的中轴线，北起东、西大街的衔接处，南至南门（迎薰门），以市楼贯穿南北，街道两旁老字号与传统名店林立，是最繁华的传统商业街。南大街有很好的旅游资源，其文物古迹、历史建筑保存完好，有保留比较完整的传统格局和独特的历史风貌，且具备一定的规模。经济文化充满活力，对原有的历史文化和社会生活有一脉相承的延续性，至今仍维持着原有的社会功能和经济文化活力。

Market Positioning 市场定位

Pingyao Ancient City has an important impact on politics, economy, culture and other aspects of China, and it features a long history, well-preserved cultural relics, historic sites, and historic buildings to reflect local and national characteristics. Thus, Pingyao South Street is positioned to rely on existing cultural relics and historic sites, historic buildings, to dig history and culture, to create a commercial district of food, housing, entertainment, sightseeing, tourism, and to showcase historic features and build a gathering place for Chinese old brands.

平遥古城在历史上对中国的政治、经济、文化等方面有过重要影响，历史悠久，文物古迹、历史建筑保存完好、丰富，能集中地反映地方、民族的风貌特征，这是平遥古城最突出的特点。因些，平遥南大街的定位是：依托现有的文物古迹、历史建筑，深入挖掘历史文化，打造成为一个集吃、住、娱、观、游于一体的商业街区，着力展示历史风貌和打造中国老字号聚集地。

Street Planning 街区规划

Pingyao Ancient City has four main streets, eight side streets, and seventy-two winding lanes. South Street is one of the four main streets. It is not only the central axis of the ancient city but also the backbone of commercial prosperity. It's a north-south street, facing the South Gate, and parallel to east walls in the east and Shaxiang Street in the west.

　　平遥古城有"四大街、八小街、七十二条蜿蜒巷"之说，南大街是四大街之一，它不仅是平遥古城的中轴线，还是平遥古城商业繁荣的主心骨。它是正对南门的南北向大街，与东面的东城墙和西面的沙巷街不但平行，而且距离相等。古代"寻龙点穴"的"金外"就在这条街上，"金片占楼"横跨街心，南大街也是这座古城的脊梁。

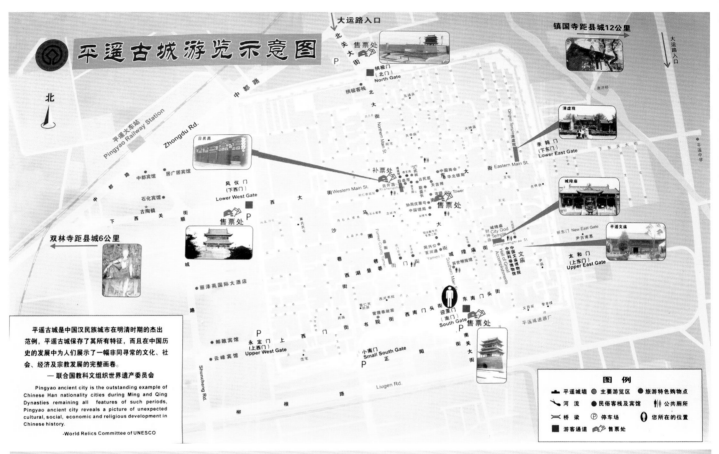

Street Design Features 街区设计特色

Pingyao South Street, known as the "Chinese Wall Street", is divided into South Street and North Street by the Municipal Building. The basic style of shop buildings connecting with court is the same, but not identical. The great shops are delicately designed with exquisite carving and painting techniques, full of beauty under a long history. The Municipal Building was re-built in the Kangxi Period of Qing Dynasty. Its initial function was managing the entire market, but now it is as the city center and the commanding height. Also known as "Gold-well Building" because of a gold-water well in the south, the 18.5 m-high building has three eaves and a gable and hip roof with colored glazed tiles, which creates patchworks of "喜" in the south and "寿" in the north. Being dignified, beautiful and richly decorated, the building stands in the city over the centuries, overlooking the magnificent mountains and rivers, and the bustling city street.

平遥古城南大街整体上保留了明清时期的街区风貌，全长690米，汇集大小古店铺78家，被誉为"中国华尔街"。早在19世纪乃至更早的时期，南大街的商业就十分兴旺，街市繁华，曾是晋商的发祥地之一。整条大街由市楼分隔为南、北街，商铺建筑及辅院相连的基本格调一致，但又绝不雷同。高大宽绰的清代铺面建筑风格考究，雕刻与绘画手法细腻精致，历经百年而古风犹存。市楼重建于清康熙年间，原起着管理整个市场的作用，如今是全城的中心和制高点。市楼原来是一方盛景，因楼南有一口井，"井内水色如金"，故又称"金井楼"，三重檐歇山式楼阁，通高18.5米，屋顶施彩色琉璃瓦，并相间拼凑为南"喜"北"寿"字样，吻兽、仙人烧造精巧，工艺水平极高。整座市楼端庄秀美，装饰富丽堂皇，几百年来，雄踞城中，远眺山河之壮丽，俯临街市之繁华。

Modern Riverside Scene at Qingming Festival
现代清明上河图

Major Commercial Activities 主要商业业态

There are lots of stores and workshops on Pingyao South Street. Restaurants, grain stores, willow baskets shop and ceramic shops are built everywhere. The food shop "Cuicheng Hailu" with high reputation is still flourishing. The modern public building — cinema has replaced the ancient Erlangmiao. There are not many boutiques, but more assortments in the paper workshops. New industries emerge in the former site of Kuitai Smoke Shop. Primitive shops with a variety of industries create the prosperity of the street.

平遥南大街中商店、作坊林立，酒家、粮店、柳箩店、陶瓷店在这一带为数众多；久享盛誉的老字号——翠成海炉食铺如今生意依旧兴隆；现代公共建筑——电影院取代了古代奉祀风俗神的二郎庙；在许多纸扎作坊里，虽少见精品，却多了许多花样；盛极一时的魁泰烟店旧址进驻了新兴的行业。一间间古朴的铺面，多种贴近人们生活的行业，共同织成了南大街繁荣的景象。

Chang Yi Feng 昌颐丰

Chang Yi Feng, founded in 1990, is connected with important streets, and close to many attractions. It is a hotel with Ming and Qing dynasties style. The interior decoration is simple and elegant. There are more than 40 rooms of various types with comfortable courtyards.

　　昌颐丰创建于1990年，位于明清一条街与衙门街的交会处北侧，步行可至百川通票号旧址、藏报博物馆、市楼、真君府、平遥县衙和平遥古城（北门）等诸多景点。客栈是古色古香的明清建筑，室内装修舒适温馨，具有古朴典雅的风格，院落环境优雅，拥有客房40余间，有土炕房、标准间和三人间等多种房型。

Bai Chuan Tong 百川通

Bai Chuan Tong is one of the most famous exchange shops of Qing Dynasty, and located in the west of Pingyao South Street. Its south-north façade leans the shops.

The former site is strictly-organized, symmetrical and enclosed with the commercial and residential functions. The whole structure is well-proportioned with easy courtyard and level access. The front and back three courtyards with the side courtyard cover an area of more than 1,300 square meters. The layout of the courtyard is arranged as the style of northern residence. There are 50 rooms with different sizes.

百川通票号旧址坐落于南大街路西，坐西朝东，南北侧与铺面相靠，始创于清咸丰十年（1860年），创始人渠源祯，至"民国"七年歇业，是清朝最有名的票号之一。

百川通旧址整体布局严谨、对称、封闭，兼有商业与民居的功能，结构错落有致，院院串通，房舍上下可达。前后三进院带偏院，占地1 300余平方米，院落布置成北方典型的民居式，有大小厅房50间。

Tongxinggong Escort 同兴公镖局

Tongxinggong Escort was founded in 1855 by a national martial arts master and developed by his son. There is still a yellow-blue plaque in the escort awarded by Empress Dowager Cixi, which makes people remember its glorious history.

同兴公镖局创于清咸丰五年（1855年），创始人是当时名扬京城、威震全国的武林大师王正清，其子王树茂尽得其真传并超越父亲，因而镖局在创立之初就全国闻名。至今镖局内还挂有当年慈禧钦赐给二代掌柜王树茂的一块黄蓝相间的"奉旨议叙"匾额，显示出它当时的风光。

Wei Sheng Chang 蔚盛长

Wei Sheng Chang is an exchange shop founded in 1826, and out of business in 1916. It's one of the five "Wei" affiliates, and also known as "Emperor Guangxu Hostel" because Empress Dowager Cixi and Emperor Guangxu had lived there. The shop sits towards the east and covers an area of about 260 square meters, and has 16 rooms. The front façade is on the street side.

蔚盛长票号成立于清道光六年（1826年），于1916年歇业，是平遥"蔚"字五联号之一。因慈禧太后和光绪皇帝曾下榻蔚盛长，因而又名"光绪皇帝下榻处"或"光绪客栈"。蔚盛长票号旧址坐西朝东，一进院，占地面积约260平方米，有房屋16间。铺面临街，面阔3间，进深两间。

Xie Tong Qing 协同庆

Xie Tong Qing is located in the west of Pingyao South Street, and founded in 1856. The initial investment was not too much, however, after a short period it developed into a famous money shop in China at that time; it was one of the ten money shops in Pingyao.

The former site includes six courtyards and underground golden vaults with an area of more than 3,000 square meters, and strict layout, specially-designed shape. The structure is solemn and stable, without losing the traditional architectural style and functional role of money shops. There are four courtyards; the front ands middle courtyards sit towards the east; and the back courtyard sits towards the south. The five halls work as shops, dividing the space into two levels.

协同庆钱庄旧址位于南大街路西，北临长升源字号，建立于清咸丰六年（1856年），财东为榆次聂店村王家和平遥王智村米家，初期投资3.6万两白银，短短数年就发展成为当时中国著名的票号，"民国"二年歇业，是平遥十大票号之一。

协同庆旧址由6个院落及地下金库组成，占地3 000余平方米，布局严谨，上下错落，搭接结构紧凑，平面呈"厂"字形，结构既庄重又稳定，同时又不失传统建筑风格和钱庄的功能。前后四进院，前院、中院坐西向东，后院坐北朝南。门厅5间兼作铺面，中间用木板分隔为上下两层，双坡硬山瓦顶。过厅5间，采用硬山顶式瓦顶。其余厢房、正房、耳房共25间，结构一般。整户院落，结构别致。

China Escort Museum 中国镖局博物馆

China Escort Museum is located in the middle of Pingyao South Street, and sits towards the east with an area of 2,000 square meters. The museum shows its development history and the interesting stories about top ten escorts in Ming and Qing dynasties. It also shows their daily things and arena rules.

中国镖局博物馆位于平遥南大街中部，坐西朝东，总面积2 000平方米。馆内主要展示了中国镖局的发展史以及明清时期中国有名的十大镖局、十大镖师和走镖过程中的逸事趣闻，展示了当时古城镖师曾用之物，镖箱、镖车、轿车、兵器和生活中用过的烟具、烟桌、家具等以及他们的生活习惯和江湖行规。

Yamen Hostel 衙门官舍

Yamen Hostel is located in the center of Pingyao Ancient City, and separated from the county government museum by a wall. It was built in 1591, and repaired for Qing Emperor Kangxi. Some famous people in the history had lived there. There are three doors and two courtyards, with large-size attached courtyards in the east and west.

衙门官舍位于平遥古城中心，与县衙博物馆仅一墙之隔，古时是县太爷招待上下级官员的场所。官舍始建于明万历十九年（1591年），清康熙年间，该县知县为准备皇帝西行对其加以修葺，光绪初年，曾国荃、张之洞等人曾在此居住多日。住院为三进两院过道厅，东西两侧有规模较大的配院，俗称"一主两跨"。

Shuozhou Old Street, Shanxi
山西朔州老街

History 历史承袭

The Shuocheng District of Shuozhou City is an old city site that is rather earlier and comparatively intact in Shanxi Province. It is a site of strategic significance in military affairs.

According to historical record, the city was built in 557 with earth walls. Shaped in a rectangle, the old city was 1,800 meters in north-south direction and 1,600 meters from east to west. The city was of 6,800 meters in circumference and covered 2,880,000 m². So far it has stood here for more than 1,400 years.

At the beginning of Ming Dynasty, the city was rebuilt. In 1370, it was restored with bricks. The city wall at that time was 12 meters in height, above which the battlement stood 2-m-high. There were 4 barbicans, 4 turrets, 12 water towers, 4 gate towers, and 4 warning beacons. The existing city has basically retained the original city layout. Apart from the south wall and gate, west wall and gate, and the turrets, other buildings have all been destroyed.

朔州市朔城区是山西省现存的历史较早、残垣保存较完整的古城址。它雄踞雁门关外，北连内蒙古，南控雁门、偏关、宁武三关，古为边陲之要塞，既可应援大同，又能拒防全晋，是历代兵家必争之地。历史上，匈奴、突厥、回纥、鲜卑、契丹、女真、蒙古等少数民族统治者南犯取晋，多先围守朔州，而后入雁门，直取晋阳。西汉韩信于马邑叛汉降匈奴、汉朝诱灭匈奴的"马邑之谋"、隋末刘武周于马邑起事斩太守、唐武德年间唐军与突厥的马邑争夺战、清代的农民熊六起义等均发生在朔州。

据《朔州志》记载，朔州古城创建于北齐天保八年（557年），其周长4 513米，为土城夯筑城垣。北齐古城平面呈长方形，南北长1 800米，东西宽1 600米，周长6 800米，总占地面积2 880 000平方米。隋、唐、辽、金各代沿用，距今已有1 400多年的历史。

元末明初，省去西北城垣之半，并利用东南隅北齐城墙重建朔州城，明洪武三年（1370年）包砖。当时城墙高12米，堞高2米，总高14米，周长4 000米，有瓮城4座，角楼4座，敌楼12座，门楼4座，烟墩4座。四门东曰文德门、西曰武定门、南曰承恩门、北曰镇塞门。现存城内街巷及布局基本保持原制，除南城墙及城门、西城墙及城门、瓮城保留外，其他建筑皆已毁坏。

Location 区位特征

Shuozhou City is located at the north of Shanxi Province, with its northwest adjacent to the Inner Mongolia Plateau, and south to the Yanmen Pass. It is 129 km from the urban downtown to Datong City, 200 km to Taiyuan City, and 502 km to Beijing. It enjoys very convenient traffic as well as well-developed post and telecommunication.

Founded in 1988 by the State Council as a provincial city, Shuozhou is a new-type base of energy and chemical industry in China, and it enjoys vast land, rich tourism resources as well as a plenty of historical relics. Numerous attractions around add beautiful landscapes to the city.

朔州市位于山西省北部，西北毗邻内蒙古高原，南扼雁门关隘。市区北距古城大同129千米，南至省府太原200千米，东到首都北京502千米。朔州市交通便利，邮电通信发达，运煤专线纵横交错，县乡公路四通八达，怀仁机场通航北京和上海。

朔州是1988年经国务院批准设立的省辖地级市，为全国新型的能源重化工基地，土地广阔，旅游资源丰富，文物古迹众多。周边的云冈石窟、北岳恒山悬空寺、雁门关、旧石器时代的峙峪人遗址、佛教圣地五台山、天池、万年冰洞、芦芽山自然保护区等编织成了美丽的旅游风景线。

Market Positioning 市场定位

The city was once the habitat of various ethnic groups. The communication and integration of those people created the local culture of unique characteristics.

The planning design digs deeply into the city cultural context to build local culture brand. It aims to combine local historical relics to build the place into "A City of Charm, Vigor and Life".

朔州曾是北狄、楼烦、林胡、匈奴、鲜卑、突厥（沙陀部）、回纥、契丹、女真、蒙古、满、汉等民族的杂居地，各民族之间相互沟通，彼此融合，取长补短，兼收并蓄，共同创造了极具特色而又自成体系的朔州地方文化。

街区规划设计着力挖掘古城的文化内涵，打造朔州雁北文化品牌，结合现有老城内的文物古迹把朔州营造为"魅力之城、动力之城、生活之城"。

Street Planning 街区规划

Planning Idea

1) Historical Concept: to inherit the texture of the ancient city, to protect the city layout and to reproduce the city culture; to reflect space features in Yanbei folk houses and to create living space of rich history and culture.

2) Ecological Concept: based on city green system, to create a space where man, city and nature coexist harmoniously; to introduce nature into city to build an ecological network which combines greening with streets.

3) Humanistic Idea: to create a modern community in which humanistic and artistic conception is combined with local features.

4) Development Concept: with open city structure, to create a modern living space to conform to modern life.

规划理念

1）历史理念。延续古城肌理，保护老城格局，再现古城文化，体现雁北民居空间特色，创造历史文化内涵丰富的住区空间。规划体现了传统文化：天圆地方——方正、围合、天人合一；九宫格——井字形、棋盘式；古代城郭——匠人营国。

2）生态理念。以绿脉为先导，人、城、自然和谐共生，将自然导入城市，建立绿、街相互交织的生态网络。

3）人文理念。塑造人文意境与本土特色相结合的现代化社区。

4）发展理念。开放的城市结构，塑造符合现代生活的城市生活空间。

Planning Layout

The project divides the old street organically with the four main streets and two green landscape belts. The east, west, north, and south streets are very important commercial axes. The north street is a core commercial axis. On the east side of the north street are office buildings, clubs, hotels, etc. Centered on the north street and Wenchang Pavilion are shops for leisure and shopping, while the east street is mainly for Tang Dynasty cultural entertainment. In general, the overall planning forms four axes, two rings, two centers and six zones.

Four Axes: the east and west streets are transverse axes, while the south and north streets are longitudinal axes; the other two axes are the two green landscape belts.
Two Rings: outside the city there is a ring green belt; there is another ring at the city core.
Two Centers: tourism center is formed around the Chongfu Temple, while the intersection of the four main streets forms the traditional business center.

Six Zones: there are 6 functional zones in the program. The commerce and trade zone is located at the intersection of four main streets for traditional business development. The folk activity zone lies on the south side of Chongfu Temple and the periphery of east gate, with Yu Chigong's Ancestral Hall serves for folk activities show area. The history display zone is gathered with ancient buildings of profound history and culture. The leisure green land zone not only shows the city history and cultural context, but also provides citizens with a place for leisure and cultural recreation. The traditional residential zone mainly harbors courtyards with gardens.

规划结构

通过东、西、南、北的商业步行街轴线和两条绿化景观带轴线，对老街进行有机组织分隔。东、西、南、北街为重要的商业轴线，北街为核心商业轴线，北门口区域为集中商业中心。北街东侧设办公、会馆、酒店等建筑；北街及文昌阁周围构成以休闲、购物为特色的街铺式商业中心；东街南北绿化景观带以东为文化核心轴线，形成崇福寺广场旅游观光、文化巡礼为特色的唐代文化休闲中心。总体规划形成"四轴、两环、双心、六片"。

四轴：以东街和西街为主的横向商业轴带，以南大街和北大街为主的纵向商业轴带，以及两条绿化景观带轴线。

两环：沿古城的环城绿化带、老城中心文昌阁周围环行商业步行"街中街"。

双心：以崇福寺为极核的旅游观光中心，以及在东、西、南、北街的交会处形成的传统商业中心区域。

六片：即6个功能区，分别为商贸活动区、民俗文化区、历史展示区、绿化休闲区、传统居住区和特色居住区。商贸活动区：主要集中在东街、西街、北大街和南街形成的交叉区，为传统商业发展的集中区。民俗文化区：主要集中在崇福寺南侧和东门外围，结合尉迟恭祠堂作为民俗活动的展示区域。历史展示区：以崇福寺、文庙、县衙和尉迟恭祠堂为历史文化的展示区。绿化休闲区：集绿化、休闲、文娱于一体，贯穿文庙、崇福寺、尉迟恭祠堂，既展示历史文化脉络，又为市民提供文娱休闲场所。传统居住区：是指经改造后形成的有雁北特色的传统居住社区。特色居住区：为多进院式花园豪宅。

Functional Space — User-friendly

Squares are arranged according to each functional zone's layout. They are generally located at the ends or crossing points of landscape axes and space sequences. They are very important landscape nodes of the block. Moreover, the architectural height of commercial spaces is harmonious with the overall space height of the old city. It reflects "user-friendly" concept in environment design and combines road, parking, business service, and supporting facilities in an organic composition, forming colorful atmosphere of lively breath.

Silhouette — Sense of Sequence

Commercial buildings along the street are mostly two-storey buildings, and only few are three-storey buildings. They are lower and lower from border to the center, with only few buildings such as Wenchang Pavilion and Drum Tower standing high, forming a staggering skyline.

Buildings — Fresh and Elegant

Commercial buildings, courtyard residences and clubs are mainly designed with style of Ming and Qing dynasties, while historical relics show Tang Dynasty style. Common residences, large shopping malls and office buildings simplify traditional architectural elements for innovation. At the same time, they incorporate modern design techniques. As a result, the entire street block is harmonious in styles, and it also fully satisfies people's demand for modern functions. Architectural features are shown as follows:

1) Historical Appearance & Modern Functional Interior: the architectural appearance retains traditional brick walls and tile roof, while interior of each building customizes designs according to modern people's lifestyle, pace and emotions to create modern leisure business and living environment.

2) Architectural Style in Search of Fresh and Elegant Cultural Taste: the buildings in the program adopt plain black brick and grey tile as main materials to combine tradition with modernity. They also inherit the essence of Chinese architecture. All buildings are arranged in a clear hierarchy with well-proportioned façade silhouettes. They not only form simple and bright modern style on the whole, but highlight the cultural deposits of traditional Chinese courtyard residences.

3) Enclosed Courtyard Space: inner courtyard layout is adopted and atrium is designed at the center of the building to offer residents a strong sense of belonging to the land and space in which they live. The street features tall gate towers, grey architectural appearances, and grey roofs. As for interior space structure, inner courtyard is enclosed by tall walls and houses, providing strong privacy. With traditional courtyard space highlighted, the building fully reflects the beauty of traditional architectural space. The combination of traditional space and modern functions and culture forms unique style and features.

功能空间——以人为本

根据各个功能区的布局安排相应广场，广场一般位于景观轴线以及空间序列的端点和交叉点上，是街区重要的景观节点。商业空间在高度上一方面与老城整体空间高度相协调，另一方面在近人尺度的环境设计上体现"以人为本"的思想，结合道路、停车设施、商业服务设施以及配套设施的有机构成，形成丰富多彩、充满动感的氛围。

轮廓线——序列感

沿街商业建筑主要为两层，局部为3层，高度从四周向中心逐渐降低。文昌阁和钟鼓楼为街区中心几个突出的高点，形成天际轮廓线。整个横向天际线由东向西为"东城口—牌坊—文昌阁—鼓楼"序列，由南向北为"南城门—文昌阁—钟楼"序列以及崇福寺建筑空间轮廓线序列。

建筑——清新高雅

沿街商业建筑、合院住宅、会所等以明清风格为主，古迹、寺庙如钟鼓楼、文昌阁、文庙、崇福寺、尉迟恭祠、民俗博物馆等以唐代风格为主。普通住宅、大型商业、办公等建筑对传统建筑元素加以简化创新，同时融入现代处理手法，既保持了整个街区风格的和谐，又满足了现代功能的需要。总体具有以下特点。

1）外部历史风貌，内部现代功能：外表保留传统的砖墙、屋瓦，而每座建筑的内部，则按照现代人的生活方式、生活节奏、情感世界量身定做，营造现代休闲商业和居住环境。

2）建筑风格追求清新高雅的文化品位：以朴实无华的青砖灰瓦作为建筑的主调，将传统与现代融合，继承中式建筑的理念精髓。建筑前后进退有序，天际线高低错落，在整体上既形成了简洁明快的现代风格，又突出了四合院的文化积淀。

3）院落围合空间：采用内院式空间布局，住宅中心设有天井，使住户对其所处的土地和空间产生强烈的归属感。街区享有高大的门楼、灰色的建筑外观、灰色的屋顶。在内部空间构造上，采用传统建筑的围合方式，宅内设置内庭院。院子用高墙或房屋围合起来，私密性强。整体上强化传统的院落空间，充分展示传统建筑空间的美感，并使之在与现代的使用功能及文化的融合中，产生独具特色的风貌。

钟楼

Major Commercial Activities   主要商业业态

1) North Gate Zone: there are large shopping malls, sunken squares, and indoor commercial streets with sunshine corridors. The large scale, rich space, complete commercial activities, and smooth circulation contribute to an updated product.
2) North Street: rich of high-end restaurants, hotels, business clubs and fitness centers.
3) Perimeter Zone of Wenchang Pavilion: it includes arcade commercial streets and open small squares; and covers food & beverage, tea show, boutique shopping, featured theme goods.
4) East Street: it includes traditional cultural relics such as Chongfu Temple, Yu Chigong's Ancestral Hall, museums, and so on. The major business activities are centered on folk culture.
5) Inner Ring from the East Gate to the North: there are lots of food wholesale markets.
6) West Street & South Street: mainly for featured theme commerce and street supporting business.

　　1）北门区域：主要为大型商场、下沉式广场以及阳光走廊室内步行街，规模大、空间丰富、业态完整、动线流畅，是街区商业的升级换代产品。

　　2）北街沿线：以高档餐饮店、酒店、商务会所、休闲健身俱乐部为主。

　　3）文昌阁周边区域：包括骑楼式步行街、开放式小广场步行街，以餐饮、茶秀、精品购物、特色主题商品为主。

　　4）东街沿线：包括崇福寺、寺前广场、尉迟恭祠堂、博物馆等传统文化场所。东街的商业以民俗文化为主，商品包括现代工艺品、文物复制品、名人字画、手工制品、地方特色物品、图书音像等。

　　5）东门内沿环城马道向北：主要为副食品批发市场，交易活跃，商户生意兴隆。

　　6）西街、南街沿线：主要为特色主题商业和街区配套商业。

Qibao Old Street, Shanghai
上海七宝老街

History 历史承袭

Shanghai Qibao Old Street that formed in later Han Dynasty, expanded in the early Song Dynasty, and achieved prosperity in Ming and Qing dynasties and has over 1,000 years of history. This old street is located at Qibao Ancient Town of Minhang District, Shanghai. As far back as the Ming and Qing dynasties, it was famous on the two banks of Huangpu River and Wusong River for its cloth, yarn, wine, carpentry, and water transport. After renovation, it has become one of Shanghai's sceneries.

上海七宝老街形成于后汉时期，发展于宋初，明清时达到繁盛，至今已有1 000多年历史。七宝老街位于上海市闵行区七宝古镇。七宝古镇是典型的城中之镇，因传说中的七件宝物——金字莲花经、神树、氽来钟、飞来佛、金鸡、玉筷、玉斧而得名。早在明清时期，七宝老街就以布、纱、酒、木器、水运而闻名于黄浦、吴淞两江地区，老街经过重新整修后，成为上海市的胜景之一。

Location 区位特征

The Qibao Old Street is located at the southwest of Shanghai suburb. It is an ancient town of Taihu Basin and a typical town within a city. It borders the Qingnian Highway on the north and links the Baonan Road to the south. The Xinzhen Road is to its east and the Qishen Road to its west. There are about 140,000 residents within the street, and the site area is 21.3 square kilometers.

七宝老街位于上海市西南郊，属江南太湖流域的千年古镇，是典型的城中之镇。它北临青年公路，南接宝南路，东倚新镇路，西靠七莘路。街内人口约14万，占地21.3平方千米。

Market Positioning 市场定位

After renovation as well as many years of development, it has formed a unique market-oriented bustling street which integrates leisure, dining, tourism and shopping as whole.

上海七宝老街自整修后，经过多年发展，已形成了独特的市场定位，成为集休闲、美食、旅游、购物于一体的繁华街市。

Street Planning 街区规划

The planning of the street can be summarized as a "丰"-shaped layout, with the streets running from north to south and lanes spreading from east to west. The entire area is divided by Puhui River, with three river bridges linking the south and north streets where personalized business activities show prosperity. The east and west parts of the street are linked by Beiguo Lane, Xujia Lane, Jiuqu Lane and so on.

七宝老街街区整体规划可用"街分南北，巷串东西，呈'非'字形格局"来概括，即老街以蒲汇河为界，将街区分为南大街和北大街，南、北大街由三座跨蒲汇河的桥梁串联，各自形成独特的商业业态，东西方向由北国弄、徐家弄、九曲弄等巷道连接。

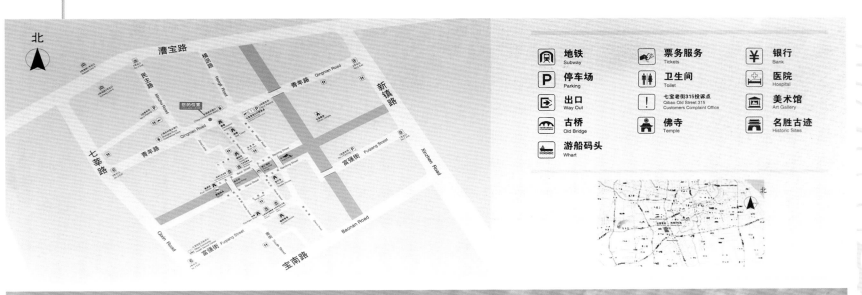

Street Design Features 街区设计特色

The Qibao Old Street has over thousand years of history. After renovation, it still shows two-storey Ming and Qing dynasties buildings built by bricks and timber, with plaster walls, black roof tiles and dark red wooden doors and windows. Narrow and deep ancient streets and lanes that were paved with black bricks are remains of Song Dynasty. Three bridges over the Puhui River and the river itself form unique Jiangnan water town scenery with water running beneath the bridges. Strolling along the old street, you will feel the profound and dignified traditional culture.

七宝老街拥有千年历史，整修后的老街两旁多是砖木混合结构的两层仿明清式建筑，粉墙黛瓦，配有深红色的木质门窗。宋代遗存的狭长的老街深巷里铺着青砖石板，蒲汇河与河上的三座小桥构成了江南小桥流水的景色。漫步老街，感受到的是传统文化的悠长与凝重。

Major Commercial Activities 主要商业业态

As far back as the Ming and Qing dynasties, the Qibao Old Street is famous on the two banks of Huangpu River and Wusong River for its cloth, yarn, wine carpentry and so on. So far, the main business activities here have transformed to special snacks, tourism crafts, antiques and paintings. The south street is for special snacks and the north street is for tourism crafts, antiques and paintings.

The special snacks here enjoy very good reputation, while excellent crafts, antiques and paintings are often found in Xunbao House for the latest stylish gifts.

　　早在明清时期，七宝老街就以布、纱、酒、木器等商品名重黄浦、吴淞两江地区，发展至今，老街的主要商业业态以特色小吃和旅游工艺品、古玩字画为主，主要分布在两大区域——南大街和北大街。南大街以特色小吃为主，北大街以旅游工艺品、古玩字画为主。

　　在特色小吃方面，七宝老街上的羊肉、糟肉、方糕、肉粽远近闻名。而在工艺品、古玩字画上，离不开古街中时尚礼品集中量贩地——寻宝楼。楼内经营最新、最时尚的礼品，主要风格类别分为民族类、创意类、家具类、文化类、美饰类，备受现代消费者关注。

品牌商铺展示

Food and Beverage 餐饮类

Qibao Old Restaurant

Qibao Old Restaurant is a time-honored dining brand,which is located at the north street adjacent to the Puhuitang Bridge. The building imitates architectural style in Ming and Qing dynasties, with rockeries and water flow through every household, giving the feeling of living in a garden. Chairs and tables are also of Ming and Qing dynasties styles, and windows are framed with wooden frames. The interior hue is dark but clean. The restaurant offers authentic Shanghai cuisine and specialties including Qibao fish head soup, roasted meat, gluten, boiled mutton and so on.

七宝老饭店

　　七宝老饭店是七宝古镇中有名的老字号，位于北大街上，邻近蒲汇塘桥，门前小桥流水，塘中鱼儿游乐。饭店主体是仿明清式的建筑，大门外修有假山流水，给人一种园林的感觉。内部桌椅是明清式家具的风格，镶着木框的窗户，整体颜色偏深，却感觉很干净。主要菜式是本帮菜，特色菜有七宝鱼头汤、红烧肉、鱼面筋、白切羊肉等。

Four Seasons Steamed Dumpling Shop

Steamed dumpling is an indigenous cuisine in China. Four Seasons Steamed Dumpling Shop, located at the Northeast Street closed to the Puhui River, is a dark red archaized building. All steamed dumplings in the shop are freshly cooked, and the menu here changes alternatively according to the seasons. Suzhou opera is also put on the stage here from time to time.

四季烧麦馆

烧麦是中国土生土长的一种美食，是以烫面为皮裹馅上笼蒸熟的面食小吃。烧麦顶端蓬松束折如花，形如石榴，洁白晶莹，馅多皮薄，清香可口。

四季烧麦馆位于老街北东街，依着蒲汇河，主体是深红的仿古建筑。馆内烧麦都是现做的，菜谱按季节制订，随季节变化而变化。馆内时常会有评弹演出。

Tangqiao Restaurant

Tangqiao Restaurant is located at the Southeast Street and overlooks the Four Seasons Steamed Dumpling Shop across the Puhui River. As a restaurant as well as a tea house, it offers authentic Shanghai cuisine. The restaurant enjoys prominent location as well as excellent Jiangnan water town scenery. The red lanterns hung on the roof eaves lighten the Puhui River as evening approaches to create romantic atmosphere here.

塘桥饭店

塘桥饭店位于南东街，与四季烧麦馆隔河而望，是七宝老街有名的饭店之一。塘桥饭店既是饭店，也是茶楼，菜式是本帮菜，味道正宗。饭店的地理位置很好，依偎蒲汇塘桥，可以欣赏江南小桥流水的特色景观。沿河的外檐上挂着一排灯笼，夜晚时分，红红的灯笼映照着蒲汇河，营造出浪漫的氛围。

Chengyue Tea House

Chengyue Tea House is located at the No.86 Northwest Street nearby the Puhui River. The building features a Ming and Qing dynasties style appearance, while its interior retains an old house flavor, with ceiling-to-floor glass windows bringing in ample daylight as well as prosperous business activities and Jiangnan water town scenery, offering guests leisure and peace.

诚悦茶楼

诚悦茶楼位于北西街86号，紧临蒲汇河，茶楼外观是明清风格，内部装修保留着老屋的风格，落地的玻璃窗使室内很敞亮，透过玻璃窗向外看，对岸的老屋、茶社、酒肆、小桥流水的景观尽收眼底，为茶客带来一份闲适与安宁。

Crafts 工艺类

Longquan Sword

Longquan Sword, originally called Seven Stars Longyuan Sword, is ranked fifth famous sword in ancient China. Legend says that it is a noble sword. The price of the craft Longquan swords in the old street varies in different shops. Cheap craft swords are offered to visitors for appreciation.

龙泉宝剑

龙泉宝剑是中国古代十大名剑之一，排名第五，相传为欧冶子和干将两位铸剑大师联手铸造，原名七星龙渊，唐代时因避高祖李渊讳，便改"渊"为"泉"，即"七星龙泉"，简称龙泉剑。

龙泉宝剑是诚信高洁之剑，这与春秋名将伍子胥有关。当初伍子胥为奸臣所害，逃至长江之滨，为渔丈人所救，伍子胥赠之以祖传宝剑——七星龙渊，并嘱咐千万不要泄露他的行踪，渔丈人说搭救伍子胥只因其为忠良，并不图报，而今伍子胥却怀疑他贪利少信，于是渔丈人以此剑自刎，以示高洁。

七宝老街上的龙泉宝剑店里面卖的宝剑价格差距较大，价低的宝剑可供游人拿在手中鉴赏。

The Memorial Hall of Artist Zhang Chongren 张充仁纪念馆

The Memorial Hall of Artist Zhang Chongren is located at one of the ends of the Puhuitang Bridge, covering 600 square meters. Launched in 2003, the hall has standed here for ten years. The building features plaster walls, black roof tiles, eaves and carved lattice works, a typical Jiangnan water town courtyard of Ming and Qing dynasties architectural style. There are six exhibition zones in the hall, and more than 20 statues created by Zhang Chongren on display.

张充仁纪念馆位于蒲汇塘桥堍，馆舍面积600平方米，于2003年开馆，至今已有10年。纪念馆整体粉墙黛瓦，飞檐雕棂，是一座明清风格的江南庭院。馆内按张充仁先生生平设置有"艺坛起步""东方英才""画室生涯""雕塑春秋""晚霞绚烂""德艺留馨"6个展区，展出张充仁的20多件雕塑代表作品。

Tongli, Suzhou
苏州同里

History 历史承袭

Tongli Town was built in Song Dynasty. It is the earliest and first culture relic protection town in Jiangsu.

According to historical record, Tongli had been an important town in Wuzhong since Song Dynasty. Because boating was the only means to connect with the outside world, it rarely suffered from wars. Therefore, it was an ideal place for the rich to keep away from social upheavals.

　　同里镇始建于宋代，至今已有1 000多年历史。同里旧称"富土"，唐初改为"铜里"，宋时将旧名拆字为"同里"，沿用至今，是江苏省最早（1982年）、也是唯一将全镇作为文物保护单位的古镇。

　　据清嘉庆《同里志》记载，从宋代起，同里已是吴中重镇，由于它与外界只通舟楫，很少遭受兵乱之灾，便成为富绅豪商避乱安居的理想之地。不难看出，同里名字的变更取决于当地人含而不露的传统观念和源远流长的历史文化。

Location 区位特征

Tongli belongs to Wujiang District, Suzhou, Jiangsu Province. It is located at the bank of Taihu Lake and the Beijing-Hangzhou Grand Canal. It is next to Suzhou City on the north and adjacent to Kunshan on the east. It is right beside the location of city government. It is near to Shanghai, Hangzhou and Suzhou. It is a Golden Delta of Jiangsu, Zhejiang and Shanghai. It is surrounded by five lakes and is divided into seven islands by braided stream.

　　同里隶属江苏省苏州市吴江区，位于太湖之滨、京杭大运河畔，北部连接苏州市区，东侧与昆山市相邻，紧靠市政府所在地。它紧依上海、苏州、杭州三大著名城市，地处江苏、浙江、上海两省一市交会的金三角地区，是中国沿海和长江三角洲对外开放的中心区域。它距苏州市18千米，距上海80千米，被5个湖泊环抱，由网状河流将镇区分割成7个岛。

Market Positioning 市场定位

Short-term positioning (2005 to 2010): taking the whole East China as core market. Making use of weekend tourist market and focally developing sightseeing tourism.

Medium-term positioning (2010 to 2020): taking Yangtze River delta, Pearl River delta and Beijing-Tianjin region as core market. Making use of weekend and holiday tourism, and giving priority to sightseeing tourism combining recreational tourism and business tourism.

Long-term positioning (after 2020): based on the whole East China, facing domestic market and developing international market. Creating a new type complex which integrates tradition and modern, combines customs and fashion, and connects meditation and exploration.

　　近期市场定位（2005—2010年）：以南京、苏州、上海、杭嘉湖所在的整个华东地区为核心市场；把握周末旅游市场，重点发展水乡古镇观光旅游。

　　中期市场定位（2010—2020年）：以长江三角洲、珠江三角洲、京津地区为核心市场，把握周末旅游、假期旅游市场，以古镇观光旅游为主，兼有休闲度假旅游、商务旅游等。

　　远期市场定位（2020年后）：立足整个华东地区，面向国内市场，开拓国际市场，形成"传统与现代的交融、民俗与时尚的握手、怀古与探奇的拥抱"的新型水乡综合体。

This town is divided into four scenic zones. They are Historical and Cultural Scenic Zone, Xiaodian Lake Forest Park Scenic Zone, Jiuli Lake Sport Leisure Scenic Zone and Nanxing Lake Rural Holiday Scenic Zone.

Garden of Seclusion and Meditations, Former Residence of Chen Qubing, Jiayin Hall, Tianfang Building and so on are restored successively. According to the general planning and protection principal, main commercial streets and buildings along the tourist route are remolded in ancient style. Wires of telephone, television and broadcast are covered by earth, and high-rise buildings are removed. A batch of residential inns are repaired and opened. A Ming and Qing dynasties style gallery is built along the river with antique tea shops. Town-wide sewage treatment is undertaken. After Nanyuan Teahouse is restored, folk art forms such as folk rap, Jiangnan music and so on are added. A folk custom street with strong Jiangnan features is open to reproduce interesting folk crafts.

New forms like ecological healthy tour, leisure experience tour and forest exploration tour are developed to turn resources advantages into industrial advantages. For example, seven lakes including Jiuli Lake, Cheng Lake, Muzhuang Lake and so on are linked in form of ecological corridors to create a 25 square kilometers' wet land.

通过"文化游"、"水上游"、"生态游"和"休闲游"四轮驱动，以文旅结合提升品牌建设。根据规划，将在镇内形成四大景区，即同里古镇历史文化景区（包括同里湖）、肖甸湖森林公园景区、九里湖运动休闲景区和南星湖田园度假景区。

项目先后对退思园、陈去病故居、丽则女校、崇本堂、嘉荫堂、富观桥、天放楼等开展重点修复；按照"古色古香"的总体规划和"修旧如旧"的保护原则，对古镇东柰、柰隶主要商业街和旅游沿线各种房屋彻底进行仿古改造；对电话、电视、广播三线实施地埋，并对高层建筑改造移位；修缮开放一批民居客栈，开发仿古式接待处；在中川桥东侧修建百米长的明清式沿河盖街廊棚，开设古色茶道茶店，并全面启动全镇的污水处理工程；以大手笔全面恢复古镇的文化内涵和水乡神韵，打造"醇正水乡，旧时江南"的古镇品牌。"江南第一茶楼"南园茶社恢复后增添同里宣卷、江南丝竹、评弹等民间曲艺表演；耕乐堂修复后再现了古园经典的风采，还设置了"大世界吉尼斯之最厅"、"自然风格厅"、"水乡艺茶吧厅"及民俗艺术表演区等；与耕乐堂同时开放的还有富有浓郁水乡特色的民俗风情一条街，再现了红土布、捏泥人、双推磨、竹器及草制品编制等别有情趣的民间手工艺。

挖掘生态健康游、休闲体验游及森林探险游等新形式，变资源优势为产业优势。如将九里湖、澄湖及沐庄湖等7个湖泊以生态走廊的形式串联起来，形成25平方千米的大型湿地。同时，将肖甸湖现有的50公顷亩平原森林资源扩展为肖甸湖休闲公园，设金色沙滩区、激情游乐区、体育运动区、森林探险区，开发热带植物园、滑雪滑冰、人工沙滩、江地世界等旅游项目。

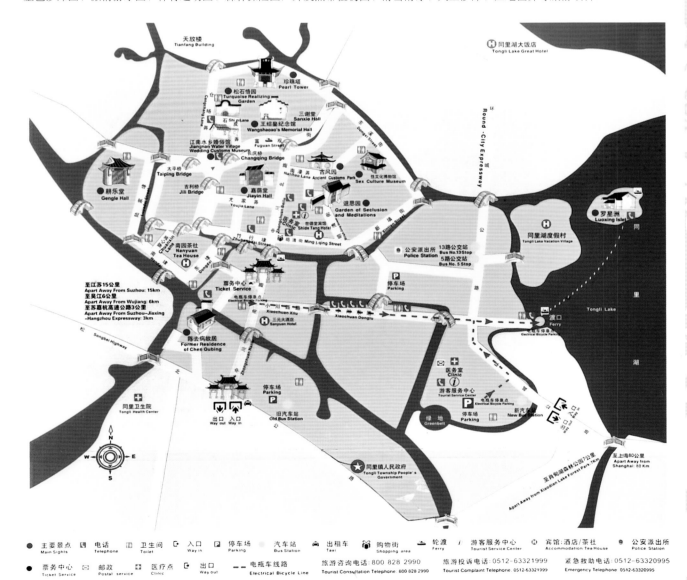

Street Design Features 街区设计特色

Organizing the scenic zones according to water system: the design smartly integrates water, roads, bridges, residences and gardens to construct a unique appearance of Tongli Old Town. The layout of bridges, streams and dwellings wins a praise of "oriental small Venice".

Numerous old houses: most of the old houses are built in Ming and Qing dynasties. Warped roof-angles, Zouma towers, brick carvings and other building details made these houses just like antique arts which survived after wind and rain.

Numerous small bridges: there are more than 40 bridges of different sizes in the town. They are mostly built after Song Dynasty. This town forms unique scenery of bridges over streams and dwellings beside streams.

Numerous lanes and alleys: between streets, there are many lanes and alleys. All these lanes and alleys are narrow and long and they will make sounds when people walking on them. Some lanes and alleys are equipped with dykes and visitors can cross the streams from one side to the other.

Buildings are built right beside water: because the town is surrounded by five lakes, almost all dwellers are living next to water. For convenience of washing, all houses are built with stone stairs at the side facing the water. There are also attics with buckets stretching out to river surface.

Exquisite brick carvings on dense buildings: there are dense buildings of big families in Tongli. More than forty buildings among them are in good keep. Brick carving is a great feature of residence in Tongli. Existing brick carvings are mostly on gate towers, screen walls and crestings of old residences and gardens, especially on gate towers.

环绕水做文章，因水成街，因水成路，因水成市，因水成园。巧妙而自然地把水、路、桥、民居、园林等融为一体，构成了古镇同里特有的水乡风貌。镇内家家临水，户户通舟；明清民居鳞次栉比；宋、元、明、清各个时期的桥保存完好。以小桥、流水、人家的格局赢得"东方小威尼斯"的美誉。

老房子多。同里"老房子"大多建于明清时期，呈脊角高翘的房屋原貌。加上走马楼、砖雕门楼、明瓦窗、过街楼等，远远望去，一组古老建筑好像是一件可以让人长久回味的古老艺术品，在沧桑风雨中，兀然矗立。

水乡小桥多。因水多，故桥也多，镇内共有大小桥梁40多座，大多建于宋代以后，著名的有建于南宋的思本桥和建于元代的富观桥。镇内自成水网，形成"水巷小桥多，人家尽枕河。柳桥通水市，河港入湖田"的独特景观。

里弄多。在街道与街道之间，里弄较多，如尤家弄、串心弄、同泰弄、西弄、仓间弄等。这些里弄又细又长，如鱼行街的串心弄长达300余米，行走时脚下会发出"哐哐"的声响；富观街附近的仓场弄，自南向北通达水河两岸，弄堂仅容一人行走，故称"一人弄"，穿过弄内人家便上河桥。还有一些里弄则常常横穿一个圩头，可以从河的这边一直走到另一边。

建筑贴水而筑，临水而建。因五湖环绕于外、一镇包含于中，因此镇上的民居几乎都择水而居，为洗涮方便，镇内家家户户都在临水的一面建有石阶，作为水河桥，既简单又实用。也有人家搭建了伸向河面的小阁楼，并专门备好吊桶，随时可以取水。

楼宇稠密，砖雕之最。同里名门望族多，楼宇稠密，粉墙黛瓦的深宅大院至今保存完好的有40余处。砖雕是同里民宅的一大景观，其技法可分浮雕、深雕、透雕、堆雕等多种。现存砖雕大部分在旧宅和园林的门楼、照墙、脊饰等处，尤以大量的砖雕门楼为多。其中，以朱宅五鹤门楼最为壮观，五只雄鹤侍立盘旋，飘逸中显露出一种威严，此门楼堪称江南砖雕艺术之精品。

Major Commercial Activities 主要商业业态

In these buildings of strong traditional Jiangnan style, the planning commercial activities include bar, restaurant, entertainment, inn, shopping and so on. On the basis of modern imitated antique buildings, delicate classic decorations are harmonious with the old town across a bridge. As for decorative materials of business buildings, modern glass and antique A-alloy doors and windows are applied to satisfy modern commercial function. As for general layout of commercial projects, traditional "straight street" and "terrace" are replaced by enclosed market form of the old town.

在具有浓郁传统水乡风貌的建筑中，规划的商业业态有酒吧、餐饮、娱乐、客栈、购物等。在现代仿古建筑的基础上，采用江南民居原有的建筑尺度比例，精致的古典建筑装饰与一桥之隔的古镇形成整体协调的风貌；在商业建筑的装饰用材上，采用了现代的大玻璃和仿古铝合金门窗，以满足现代商业功能的需求；在商业项目整体布局上，突破传统商业街"直街"和"排屋"的形式，形成了古镇围合式的商市形态，并通过建筑单体的错落变化、商业分区的自然分隔使整个商市内部充满情趣。

Former Residence of Wang Shaoao 王绍鏊故居

It is also called Liugeng Hall. The houses are standing one by one, and the interior is winding with east, west and middle lanes.

王绍鏊故居亦称留耕堂，是中国民主促进会创始人之一王绍鏊的故居。屋宇鳞次栉比，内宅曲折，东、西、中三条陪弄进出，有"侯门深如海"之感。

China Tongli Film and Television Produce Base 中国同里影视摄制基地

It is cooperatively established by China Film Association, Jiangsu Film Association and Jiangsu Wujiang City Government and opened in April 18th, 1999. It is one of the ten film and television produce bases.

中国同里影视摄影摄制基地由中国电影家协会、江苏省电影家协会、江苏省吴江市人民政府合作建立，于1999年4月18日举行了揭牌仪式，为中国十大影视基地之一。

In 1980, it was listed as one of Scenic Spot of National Tai Lake Scenic Zone.
In 1982, it was listed as Provincial Culture Rrelic Protection Site.
In 1992, it was listed as Provincial Culture Relic Protection Town.
In 1992, the famous scenic spot "Garden of Seclusion and Meditations" was listed as world Cultural Heritage by UNESCO.
In 1995, it was listed in the first batch of Famous Historical and Cultural Town of Jiangsu Province.
In 2010, it was evaluated as National AAAAA Tourist attraction by National Tourism Administration.

1980年被列为国家太湖风景区景点之一。
1982年被列为省级文物保护单位。
1992年被列为省级文物保护镇。
1992年著名景点"退思园"被联合国教科文组织列入世界文化遗产。
1995年被列为江苏省首批历史文化名镇。
2010年被国家旅游局评定为国家AAAAA级旅游景区。

Zhouzhuang, Suzhou
苏州周庄

History 历史承袭

Zhouzhuang is initially named "Zhenfengli". It's reported that "Zhouzhuang" is named to remember its donator. In 1086, it was like a village. In 1127, its population started to increase. In the middle of Yuan Dynasty, a rich family named "Shen" settled in a village in the east of Zhouzhuang, who made it develop quickly to be a formal market town centered by Fuan Bridge. After the development in Ming and Qing dynasties, it became a big market town. However, its name is still "Zhenfengli". Until Kangxi period, "Zhenfengli" was changed to "Zhouzhuang".

According to the cultural relics, in 1761, a government sector moved to Zhouzhuang, who boosted its development. With its location advantage, it became a trading center of grain, silk and a variety of handicrafts. Handicrafts and commerce of Zhouzhuang developed rapidly with the most prominent products like silk, embroidery, bamboo ware, and so on. Luo Zhewen, a famous architect praised that Zhouzhuang is not only a treasure in Jiangsu Province, but also a national treasure.

周庄旧名贞丰里。据史书记载，北宋元祐年间(1086年)，周迪功郎信奉佛教，将庄田200亩（约13.3公顷）捐赠给全福寺作为庙产，百姓感其恩德，将这片田地命名为"周庄"。那时的贞丰里只是集镇的雏形，与村落相差无几。1127年，金二十相公跟随宋高宗南渡，迁居于此，人烟才逐渐稠密。元朝中期，颇有传奇色彩的江南富豪沈万三之父沈佑，由湖州南浔迁徙至周庄东面的东宅村（元末又迁至银子浜附近），因经商而逐步发达，使贞丰里出现了繁荣景象，形成了南北市河两岸以富安桥为中心的旧集镇。到了明代，镇廓扩大，向西发展至后港街福洪桥和中市街普庆桥一带，并迁肆于后港街。清代，居民更加稠密，西栅一带渐成列肆，商业中心又从后港街迁至中市街。这时该地已成为江南大镇，但仍叫贞丰里。直到清康熙初年才正式更名为周庄。

在镇郊太师淀中发掘的良渚文化遗物记载：乾隆二十六年（1761年），江苏巡抚陈文恭将原驻吴县角直镇的巡检司署移驻周庄，管辖澄湖、黄天荡、独墅湖、尹山湖和白蚬湖地区，几乎有半个县的范围。周庄由原来的小集镇迅速发展为商业大镇，西接京杭大运河、东北接浏河，变成了一个粮食、丝绸及多种手工业品的集散地和交易中心。周庄的手工业和商业得到迅猛的发展，最突出的产品有丝绸、刺绣、竹器、脚炉、白酒等。著名建筑学家罗哲文盛赞周庄"不但是江苏省的一个宝，而且是国家的一个宝"。

Location 区位特征

Zhouzhuang is located in southeast of Suzhou and southwest of Kunshan, and has the reputation of "China's first water town". It's about 45 kilometers from Suzhou and about 100 kilometers from Shanghai. There are three airports nearby: Shanghai Hongqiao Airport, Pudong Airport and Xiaoshan International Airport. There is no train; there is no direct access to highway, but ordinary roads from SuZhou-Shanghai and Suzhou-Jiaxing-Hangzhou high-speed road.

周庄位于苏州城东南、昆山西南，有"中国第一水乡"的美誉。镇为泽国，四面环水，咫尺往来，皆需舟楫。距离苏州城约45千米，距离上海约100千米。附近有上海虹桥、浦东以及萧山国际机场。周庄不通火车，暂时没有高速公路直达，从沪苏高速、苏嘉杭高速下来需走普通公路。

Market Positioning 市场定位

Zhouzhuang is one of six ancient towns in Jiangnan. With long history, rich culture, beautiful scenery, unique cultural landscape and rustic folk custom, Zhouzhuang becomes a treasure of oriental culture. As an outstanding representative of Chinese traditional culture, it is the cradle of Wu culture.

Zhouzhuang is committed to mining, promoting and passing down the excellent traditional culture. Actively exploring the cultural tourism to create "Folk Zhouzhuang, Livelihood Zhouzhuang, Culture Zhouzhuang", Zhouzhuang is becoming a window of showing Chinese culture to the world.

周庄是江南六大古镇之一。由千年历史沧桑和浓郁吴地文化孕育的周庄，以其灵秀的水乡风貌、独特的人文景观、质朴的民俗风情，成为东方文化的瑰宝。它作为中国优秀传统文化的杰出代表，是吴地文化的摇篮、江南水乡的典范。

周庄致力于优秀传统文化的挖掘、弘扬和传承，积极探索文化旅游，全力塑造"民俗周庄、生活周庄、文化周庄"，集旅游、住宅、购物、休闲度假、餐饮于一体的"中国第一水庄"正日益成为向世界展示中国文化的窗口。

While strengthening infrastructure construction, Zhouzhuang actively carries out sincerity service, and is committed to creating a policy, investment and living, and human environment, and continues to build unique new competitive advantages.

Through the highlighting of investment and the launching of modern leisure experience-based tourism supporting projects, Zhouzhuang tourism gradually develops into leisure resort tourism. Zhouzhuang is ready to take off to the world with international concept, classic water town culture, unique tour route, supporting facility and international reception capacity.

项目在加强基础设施建设的同时，积极开展诚信服务，致力于营造政策环境、投资环境、生活环境、人文环境，不断地构建独特的竞争新优势。

项目不断加大招商引资力度，富贵园、江南人家、钱龙盛市等现代休闲体验型旅游配套项目的相继推出和完善，使周庄旅游逐步向休闲度假型旅游发展；提出了打造"国际周庄"的构想，借助经典的江南水乡文化来展示优秀的中华文明，以文化的交融为切入点，把周庄推向国际；通过资源的整合，推出适宜现代体验式旅游的精品线路和项目，加大投入完善旅游配套设施和提高国际接待能力，努力把周庄建设成为国际休闲度假基地。

Zhouzhuang successfully builds two well-known brands: "Jiangnan Town Tour" and "Sensor Industry Base", and realizes the joint development of tourism and high and new technology industry.

With unique tourism resources, the development of Zhouzhuang adheres to the "protection and development" concept to vigorously develop tourism: regarding the ancient town as the basis, continually taping the cultural connotation, perfecting scenic construction, richening travel content, enhancing promotional efforts to build a tourism cultural brand.

周庄成功地走出了一条旅游业和高新技术产业两翼并行发展的新路，开创了"江南水乡古镇游"和"传感器产业基地"两个著名品牌。

周庄凭借得天独厚的水乡古镇旅游资源，坚持"保护与发展并举"的理念，大力发展旅游业。以水乡古镇为依托，不断挖掘文化内涵，完善景区建设，丰富旅游内容，强化宣传促销，努力打造"中国第一水乡"的旅游文化品牌，开创了江南水乡古镇游的先河。

Street Planning 街区规划

Zhouzhuang developed slowly until 1986 for the negative location and inconvenient transportation. Later, a particular guideline named *Zhouzhuang Master Plan & Ancient Town Protection* Plan was made: protect ancient town, build new area, develop tourism and enhance economy.

In 1997, a guideline *Zhouzhuang Area Protection Plan* raised the new evaluation system of town planning, started to research Zhouzhuang social and cultural issues, spatial structure and morphological context in different historical periods, and how to renovate important rivers and streets. The guideline adapts foreign mature historic towns planning ideas combined with Zhouzhuang features.

In 2002, due to many negative effects of tourism development, another guideline *Remediation Plan of Key Sections of Zhouzhuang Ancient Town District* was made to control the construction size and four functions: residence, tourist, cultural display and landscape leisure, to determine the layout of each business area.

In 2005, there was a new planning idea including four points:
1. View Zhouzhuang from different layers and seek coordination at regional levels.
2. Maximize the use of resources and combine spatial layout and social development.
3. Protect Zhouzhuang through a variety of means.
4. Focus on exploring Zhouzhuang's ecological advantages.
   Planning objectives are "protecting the town, respect for ecology, featured tourism and harmonious development."

1986年，由于周庄位于诸县之交，不通汽车，远离大城市，因而发展缓慢。因此，政府制定了《周庄镇总体规划及古镇保护规划》，明确提出"保护古镇，建设新区，开辟旅游，发展经济"的十六字方针。

1997年，制定《周庄古镇区保护规划》，该规划对古镇的现状提出新的评价体系，研究周庄古镇的社会与人文等问题；研究古镇不同历史时期的空间结构、形态脉络；对沿河、沿街等重点地段进行整治；专题研究市政设施配套，尤其是污水处理问题。该规划吸收了国外成熟的历史城镇的规划思路，并融合了周庄特色。

2002年，由于旅游发展带来了很多负面效应，因此制定了《周庄镇古镇区重点地段整治规划》，划定了单个文物点、古镇区、古镇区外围建设的控制范围，将用地结构整合为居住生活区、旅游服务区、文化展示区和景观休闲区4类功能区，确定了各个地段的商业布局。

2005年，新的规划思路主要有以下4点。

1. "跳出古镇看周庄"，在区域层面寻求协调；
2. 最大限度地利用好周庄各种物质资源，空间布局和社会发展相结合；
3. 通过各种手段保证古镇保护；
4. 重点发掘周庄"水乡湿地"的生态优势。

规划目标为"保护古镇、尊重生态、特色旅游、和谐发展"。

Street Design Features 街区设计特色

The spatial combinations and courtyard houses of buildings are typical for Jiangnan and Kunshan. Private garden adopts traditional gardening practices based on residence and courtyard. Multi-door houses are usually wood structure with brick gatehouse, wooden beams, small tiles, sloping roof, projecting front and rear eaves, bungalows, dormers and so on. The water system, combined with traditional green layout and space axis, unifies water green land, corner green land, room green land and courtyard green land as a whole to create a natural and comfortable green environment in city center.

The bridges and streets are connected, and the buildings are along the rivers. More than 60% of houses are Ming and Qing dynasties residential buildings, and there are nearly 100 classic houses and more than 60 classic brick gatehouses, which are all of primitive beauty. There are 14 distinctive bridges, and together with the water and people, they create a wonderful watery landscape. Four main rivers intersect with each other, and form a perfect landscape.

　　建筑具有江南和昆山地特有的空间组合方式和庭院式民居形式，私家园林以住宅、院落为基础，采用传统造园手法，其中，多进式民居一般为硬山造木结构，设有砖雕门楼、木梁柱、小青瓦、坡屋顶、前后屋檐外抛、平房和楼房屋面开老虎窗、盖明瓦等。在街区水系方面，结合传统民居以小见大的绿化布局形式与空间轴线，通过步行路线和恢复河道水系将水景绿地、转角绿地、宅间绿地、庭院绿地等贯穿为统一的整体，创造一个自然、亲切、有传统民居特色的城市中心绿化环境。

　　桥街相连，依河筑屋，全镇60%以上的民居仍为明清建筑，有近百座古典宅院和60多个砖雕门楼。周庄民居，古风犹存，最有代表性的当数沈厅、张厅。同时，周庄还保存了14座各具特色的古桥，共同勾勒了一幅美妙的"小桥、流水、人家"的水乡风景画。南北市河、后港河、油车漾河、中市河形成"井"字形，傍水筑屋，依河成街，深宅大院，重脊高檐，河埠廊坊，过街骑楼，穿竹石栏，临河水阁，一派古朴幽静的景象。

Major Commercial Activities 主要商业业态

There are three commercial activities in Zhouzhuang: accommodation and catering, characteristic farms, handicraft and souvenir. There are rich resources and choices for tourism. Hotels, hostels, restaurants and shops can all meet various needs.

Shenting Restaurant, located at Fuan Bridge, has retained the Ming and Qing dynasties style, elegant and classic. It's a local-style restaurant.

The water town has aquatic products of all seasons. Some delicious pickles have local features. Some cakes and cooked food are of various kinds.

主要商业业态是：一是以住宿、餐饮为核心的产品；二是特色农庄项目，这也是周庄着力发展现代农业、助推旅游转型升级的龙头项目；三是依托古镇丰富的文化，开发的旅游工艺品和纪念品。周庄共有中小型宾馆5家、客栈130多家、餐馆200多家、就餐位10 000多个、床位4 000余张，形成了完整的"吃、住、行、游、购、娱"一条龙旅游体系。

元末明初时期，沈万三成为江南首富，特聘名厨烹调各式佳肴，冠以"万三家宴"，其宴讲究时鲜，选料精致，色、香、味、形俱佳。位于富安桥的沈厅酒家至今仍保留着明清风貌，典雅别致，临河傍水，是极具地方风格的菜馆。

水乡周庄，珍馐水产四时不绝，其中最有名的是"蚬江三珍"——鲈鱼、白蚬子、银鱼。周庄还出产鳗鲡，"稻熟鳗鲡赛人参"这句乡谚人尽皆知，此外还有甲鱼、河虾等；江南特产的腌菜苋、青团等也深受游人喜爱，还有数不尽的糕点熟食，如芝麻糕、花生糕、胡桃糕、椒盐糕、青糕等。

Food & Beverage　餐饮类

Ancient Stage Restaurant

It's an ancient restaurant with local features and signature dishes. White concrete square column defines the outline of the building, adding a bit of modern flavor. There is a relatively wide entrance with the height of one level.

古戏台大酒楼

　　古戏台大酒楼的招牌菜有万三蹄、清蒸白水鱼、阿婆菜、美人椒排骨、大盆酸辣鱿鱼、青椒爆肥肠等。白色的混凝土方形柱体界定了建筑的轮廓，增添了些许现代的气息。

Starbucks

Founded in 1971, Starbucks is the biggest coffee multiple shop, which sells coffee and other products, like tea, paste cake, and cake. Its head office is in Seattle, Washington, USA. The coffee emerges into Zhouzhuang, takes off the modern favor and takes on a classic appearance. The glass window features the coffee, and what's more, gives the street more light and vigor.

星巴克

　　星巴克于1971年成立，是全球最大的咖啡连锁店，总部坐落在美国华盛顿州西雅图市，除咖啡外，亦有茶、馅皮饼、蛋糕等食品。星巴克褪去了现代的衣装，转身来到周庄，融入周围古朴的氛围之中。玻璃窗是星巴克的一个身份特征，同时为微暗的街道带来光亮和生气。

Lou Wai Lou Restaurant

Founded in 1848, the restaurant has a long history of more than 100 years. Two-storey structure echoes with the surrounding buildings. Dark wood covers the façade and has a good performance here. Big-size doors and windows, especially the windows on the second floor, create a transparent space, and people indoor have a good view and privacy. The lights are designed with red tone to create a

楼外楼老饭店

楼外楼老饭店创建于清道光二十八年（1848年），是一家具有一百多年悠久历史的老店，店内名厨云集，佳肴迭出，西湖醋鱼、宋嫂鱼羹、蜜汁火方等风味独具一格。饭店曾接待过西哈努克亲王等国家元首和众多国外贵宾，在国际上享有盛誉。两层楼的结构与周围的建筑相呼应，没有突兀之感。黑色原木在这里得到充分的运用，包裹着立面，有一种历史的凝重感。大尺寸的门和窗，特别是二楼的窗，创造了一个通透的空间。游人在这里既可以享受绝好的视野，同时也可根据自身需要享受秘密空间。

San Yuan Lou

The restaurant also has a long history with three levels and overall wood structure. There are front and back doors on the first level, which are at the same line and therefore promote cross-ventilation.

三元楼

三元楼始建于明嘉靖年间。建筑为三层，全为木结构，底层有前后两门直通，正面竖有直匾，上书"三元楼"；三楼塑有魁星一尊，手执金光闪闪的大笔一支。据说被魁星的大笔点中的读书人定能考中，交上好运。因此，每逢县考、府考、院考（省学政主持的考试），凡来应试的读书人，都踊跃前往祈福，楼上人来人往，络绎不绝，热闹非凡。民间流传的"魁星点斗"之说，就由此而来。乡试第一名称"解元"，会试第一名称"会元"，殿试第一名称"状元"，特取名为"三元楼"。

Paper Box Kingdom Creative Restaurant

The restaurant is a paper box kingdom with the design concept that paper can create a new life, and make the life more interesting. It consists of "Paper Box Story House", "Honey Story House", "Paper Coffee", "Paper Post Office", "Wansan Doll Museum", "No-Sugar Tribe" and "Creative Food". There are three-dimensional paper sculptures about Zhouzhuang and the world in the space, which are exquisite and water-resistant, and also can be made into colorful nightlights, dazzling and attractive. The restaurant takes advantage of paper to reproduce Zhouzhuang's attractions and concentrate the beauty of water town.

纸箱王主题创意园区

"纸箱王"是一个结合设计人、创意人的纸箱王国，用纸玩味生活、设计生活、创造生活是纸箱王的理念。园区由"纸箱故事馆""蜜蜂故事馆""纸咖啡""纸邮局""万三公仔馆""无糖部落"及"创意餐饮"组成。主题创意园区以纸再现了周庄景点，浓缩了水乡之美，在周庄古镇的人文空间里装点了创意特色，将"纸"的创意无限地延伸……园区内摆放着原材料为纸的周庄双桥、水乡渔船、牛娃放牛、聚宝盆等立体纸雕作品。此外，还有埃菲尔铁塔、比萨斜塔、帆船酒店大厦等世界知名建筑的立体纸雕作品，不但精致美观、防水耐腐，还可以做成七彩夜灯，入夜后灯光灿烂，琳琅满目。

Dengxu Abbey 澄虚道院

The abbey has a long history of more than 900 years. Since the mid-Ming Dynasty, its scale gradually increased. Until the time of Qianlong period in Qing Dynasty, it developed into a building in front and at back of which have three doors, covering an area of 1,500 square meters. It's very famous and there are some pavilions in the building.

澄虚道院俗称"圣堂",建于宋元祐年间(1086—1093年),距今已有900多年历史。自明代中期以后,道院规模日趋恢宏,清乾隆时期已形成前后三进的宏大建筑,占地1 500平方米,为吴中地区的知名道院之一。院内主要建筑有玉皇阁、文昌阁、圣帝阁等。

Nanhu Guchin Club 南湖古琴社

From the beginning of Han Dynasty, Guchin gradually became a literatus' instrument. The literati gathering is an important channel for literature and art communication. The present Nanhu Guchin Club functions as the gathering place and a place for visitors to appreciate the music and have a chat.

从汉代始，古琴渐渐成为文人的乐器，雅集也成为文艺交流的一种方式。现在的琴社既为古琴学员雅集之用，同时也是游客们欣赏古琴、闲余茶话的场所。

Shenting 沈厅

Shenting was built in 1742 with an area of 2,000 square meters. There are seven doors and five gatehouses. The axis is 100 meters, along two sides of which there are 100 different-size rooms. Shenting is divided into three parts. The front part is the water-wall door and the port for the use of boating and washing; the middle part is a wall gatehouse, tea room, main hall for guests, weddings and mourning, and meeting; the back part is a big lobby, small lobby and rooms for the living.

沈厅由沈万三后裔沈本仁于清乾隆七年(1742年)所建。建筑为七进五门楼，有大小100多间房屋，分布在100米长的中轴线两侧，占地2 000多平方米。沈厅由3部分组成：前部是水墙门、河埠，供家人停靠船只、洗涤衣物之用；中部是墙门楼、茶厅、正厅，为接送宾客、办理婚丧大事及议事之处；后部是大堂楼、小堂楼、后厅屋，为生活起居之所。

Double Bridges 双桥

The bridges are in the center of Zhouzhuang, built in Ming Dynasty. It consists of two bridges, the surfaces of which are separately horizontal and vertical and the apertures are square and circle, like an ancient key.

双桥指位于周庄中心位置的世德桥和永安桥，建于明代。两桥相连，桥面一横一竖，桥洞一方一圆，样子很像古代的钥匙，又称"钥匙桥"，因出现于旅美画家陈逸飞的油画《故乡的回忆》中而闻名。

Zhouzhuang Museum 周庄博物馆

There is a front hall and a back hall, two sides of which are rooms. There is a patio in the center. The back courtyard is full of bamboos. The museum has many cultural relics and famous paintings, photos, arts and crafts.

来到周庄博物馆，跨进门槛，依次步入前厅后堂，两边厢房夹拥，中间天井贯通。后边小院内修竹丛生，幽径生趣。馆内有镇北太史淀湖底出土的良渚文化和印纹陶文化文物，也有出自现代著名艺术家之手的名画、摄影作品和工艺美术作品等。

Zhangting 张厅

Zhangting was built in Ming Dynasty with an area of over 1,800 square meters. There are seven doors and more than 70 rooms. It's a richly ornamented building, and carved beams and painted rafters create a glorious space. There is Ming Dynasty style redwood furniture in the hall and calligraphy and painting on the wall.

　　张厅原名怡顺堂，建于明代，清初转让张姓，改为玉燕堂，俗称张厅。张厅前后七进，房屋70余间，占地1 800多平方米，雕梁画栋，金碧辉煌。厅堂内布置明式红木家具，墙上悬挂字画，一副对联尤其引人注目，上联是"轿从门前进"，下联是"船自家中过"，贴切地道出了张厅的建筑特色。

Shortlisted as a potential UNESCO World Cultural Heritage Site
Dubai International Award for Best Practices in Improving the Living Environment
UNESCO Asia-Pacific Heritage Award for Cultural Heritage Conservation
The World's Most Charming Water Town
One of China's Top 50 Places Most Worthy of Visiting According to Overseas Tourists
One of the Top 10 Scenic Spots in China as Favored by European Tourists
Elected as one of the First Group of Famous Historic and Cultural Towns in China
National Town of Environmental Beauty
Well-known Tourist Brand in China
National AAAAA Tourist Attraction
CNN One of Five Most Beautiful Water Towns

列入联合国教科文组织世界文化遗产名录预备清单

联合国迪拜国际改善居住环境最佳范例奖。
联合国亚太地区文化遗产保护杰出成就奖。
世界最佳魅力水乡。
中国最值得外国人去的五十个地方之一。
欧洲人最喜爱的中国十大景区之一。
中国首批历史文化名镇。
全国环境优美镇。
中国旅游知名品牌。
国家AAAAA级旅游景区。
美国有线电视新闻网（CNN）评出的"中国最美的五大水乡"之一。

Wuzhen, Tongxiang
桐乡乌镇

Street Background & Market Positioning

街区背景与定位

History 历史承袭

The ancient name of Wuzhen is Wudun and Wushu. Wuzhen is a plain flooded by rivers, raising like a dark hillock. In Spring and Autumn Period, Wuzhen is the border between Wu State and Yue State. Wu once arranged army to guard against Yue.

Wuzhen was divided and controlled by different parts from Qin Dynasty. In Tang Dynasty, Wuzhen was governed by Suzhou. Now, Wuzhen is governed by Tongxiang County.

乌镇古名乌墩、乌戍。乌镇是河流冲积平原，沼多淤积土，故地脉隆起高于四旷，色深而肥沃，遂有乌墩之名；春秋时期，乌镇是吴越边境，吴国在此驻兵以防越国，"乌戍"由此而来。

秦时，乌镇属会稽郡，以车溪(今市河)为界，西为乌墩，属乌程县，东为青墩，属由拳县，乌镇分而治之的局面由此开始。唐时，乌镇隶属苏州府，唐咸通十三年(872年)的《索靖明王庙碑》首次出现"乌镇"的称呼。元丰初年（1078年），已有分乌墩镇、青墩镇的记载，后为避光宗讳，改称乌镇、青镇。1950年5月，乌、青两镇合并，称乌镇，属桐乡县，直到今天。

Location 区位特征

Tongxiang is famous not only for its economic development, but for a large amount of famous and rich culture. Its tourism culture features a well-known water town and four "Golden Phoenix".

Wuzhen has a good location, in the north end of Tongxiang and connecting with two provinces and three cities. It also connects with many national roads.

桐乡不但以经济发达著称，更以名人荟萃的文化之邦享誉国内外。以闻名全国的江南水乡古镇——乌镇和四只"金凤凰"为主架的名人文化建设，成为桐乡旅游文化的特色和优势。

乌镇地处桐乡市北端，西临湖州市，北界江苏吴江县，为二省三市交会之处。陆上交通有县级公路姚震线贯穿镇区，经姚震公路可与省道盐湖公路、国道320公路、318公路、沪杭高速公路相衔接。乌镇距桐乡市区13千米，距嘉兴、湖州、吴江三市分别为27千米、45千米、60千米，距杭州、苏州均为80千米，距上海140千米。

Market Positioning 市场定位

Wuzhen is a perfect combination of authentic Jiangnan watertown and world-class leisure resort. It focuses on cultural tourism, and features arts tourism, religious tourism, shopping and leisure tourism.

乌镇的市场定位为：原汁原味的中国江南水乡与世界一流休闲体验型度假区完美结合；建设以名人文化旅游为主题，民俗民艺旅游、宗教旅游、购物休闲旅游并举的现代旅游度假名镇。

Development Concept

开发理念

The development concept is to repair the old as before and keep its true, and its principle is ecological protection and modernization. Development and protection are on the same level, the main plan of which is to govern the external environment and improve living conditions, and strive to achieve harmonious development of people, environment, nature, and cities and towns.

In the development, the protection of historic streets is upgraded, including not only the historic buildings, but also folk culture related to the water town. The ancient town should not only be developed for view, but for comfortable life, which should be sustainable.

彻底贯彻"修旧如故，以存其真"的理念，在保护过程中以"生态保护，环境第一"和现代化为原则，把外部环境的治理与居住条件的改善作为保护与开发的重要内容，努力实现人和环境、自然、城镇的和谐发展。

在开发中，把历史街区的保护提升到了更高的层次，不仅保护历史建筑，更保护与江南水乡相关的民情民俗、市井氛围和博大的民俗文化，提出了"古镇不仅是给人看，更适合人居住"的可持续发展观。

Wuzhen is the pioneer of the development and protection of Chinese historic streets on "pipeline buried, lavatories engineering, dredging, repairing the old as before, controlling excessive commercialization, protecting folk culture", which reproduces its original style and achieves the organic historical and modern integration. Wuzhen has already become a famous tourism brand and representative with high reference value.

Wuzhen Xizha is the second phase with the development aim of best quality and relaxation. It was open to the public in October 2005 after the protection, excavation, restoration of the rivers, streets, bridges, folk and others. Some traditional folk art activities are recovered, and through the historic information of local traditional handicrafts, some typical workshops are reproduced. In 2001, the 1st Wuzhen Temple Fair was held to reproduce folk festivals that disappeared for years. The above protective measures not only retain the authentic Jiangnan style, but also create a rich historical and cultural atmosphere.

　　乌镇首创"管线地埋、改厕工程、河道清淤、修旧如故、控制过度商业化、对民俗文化的挖掘保护"等中国历史街区的保护模式，这些保护措施使得乌镇的原始风貌得到了完整的重现，实现了历史和现代的有机融合，使乌镇这一默默无闻的江南小镇成为享誉中外的知名旅游品牌，"乌镇模式"成为业内人士参照的标准和典范。

　　以精品化、休闲化为发展目标的二期工程——乌镇西栅，对河道、街区、桥梁、民俗等进行保护性整理、挖掘、修复，并于2005年10月正式开放。恢复了乌镇花鼓戏、皮影戏、三跳、拳船、高杆船表演等民间传统艺术活动；对当地传统手工技艺历史资料的挖掘整理使得乌镇的蓝印花布、三白酒、刨烟坊、铜饰铺、布鞋店、竹木雕、制笔坊、糕点斋等传统作坊又出现在乌镇的大街小巷；2001年举办了"第一届乌镇香市——江南水乡狂欢节"，使香市这一消失多年的民间节庆活动重现人间，并获得了巨大的成功。在香市期间，许多传统的民俗活动，如蚕花大会、瘟元帅会、水龙大会、踏白船、地方戏汇演、高杆船、拳船表演等都得到了展示，部分活动还被保留下来，常年向游客进行表演。这些保护措施使乌镇不仅保留了原汁原味的江南水乡风情，而且还具有浓郁的历史人文气息，"不一样的乌镇"由此享誉中外。

Street Planning　街区规划

The planning starts water, electricity, road and other works, highlights group development and enhances the quality of the core. The planning also leads a new layout and enlarges the system of Wuzhen tourism area by adding ecological practices and modern agriculture and leisure travel, which can improve the content and quality; also by leading other 16 administrative villages to combine different-scale tours together and enhance visitors' happiness and satisfaction.

　　规划启动了乌镇国际旅游区水、电、路等各项工作，突出组团式开发建设，提升核心区域发展品位，引导镇区、景区协调、集聚、高效发展；把龙翔补农生态实践游和石门现代农业休闲观光游统一纳入乌镇国际旅游区体系，不断提升旅游的内涵和品位；以乌镇现有的东西栅龙头带动其他16个行政村，使大旅游与小旅游相结合，提升游客的幸福感和满意度，形成"乌镇大旅游很热闹，村里小旅游特色多"的新格局。

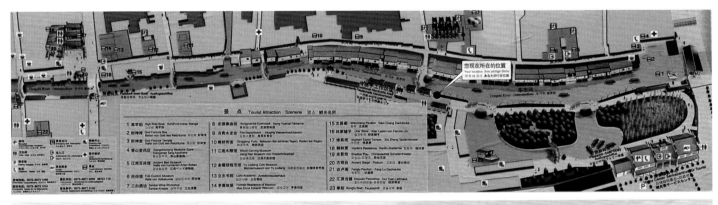

River-nets coincide with the main road and connect with bridges, making a form that the roads and rivers are like twins. This water net is connected with the Grand Canal, Taihu Lake and Wuzhen ponds, wells, to meet kinds of needs. The residences in Wuzhen, influenced by its location in history and the culture of Wu and Yue, feature multi-axis and grade.

The four river-side streets are paralleled, and water and land are adjacent. Every corner embodies a harmonious overall beauty of humanistic and natural environment. The houses are closely linked to the river, together with other ancient buildings making a unique charm of bridges, water and old houses.

Streets and houses are all along the creeks and rivers. Wuzhen is identified by "Water Pavilion", which is unique and typical. "Water Pavilion" is part of houses along the rivers that is extended above the water and fixed in the river bed with beam and board. It is the real "pillow river" with windows on three sides and a good view.

There is a bridge in every 100 feet. Bridge is an indispensable factor in Jiangnan water town. Through the long history of Wuzhen, the amount of bridges could reach more than 120. Now more than 30 bridges remain, which are rich in historical and cultural atmosphere.

河网在镇内和主干道重合，连桥成路，营造亦路亦水的形式。这个水网体系连接京杭运河、太湖和乌镇的池塘、水井，成功地解决了农作、饮用、排水、观赏、运输等水问题。在设计中，由于乌镇在历史上曾地跨两省（浙江、江苏）、三府（嘉兴、湖州、苏州）、七县（乌程、归安、崇德、桐乡、秀水、吴江、震泽），加之吴越文化的积累、沉淀，观念上受中国传统儒家文化和运河商业文化的影响，形成多轴线明确、卑尊有序的各式住宅。

双棋盘式河街平行、水陆相邻的古镇格局。作为历史古镇，乌镇无论是整个镇还是观前街，都体现着一种人文环境、自然环境和谐相处的整体美。古风犹存的东、西、南、北四条老街呈十字交叉，构成双棋盘式河街平行、水陆相邻的古镇格局。这里的民居宅屋傍河而筑，街道两旁保存有大量明清建筑，辅以河上石桥，体现了小桥、流水、古宅的江南古镇风韵。

街道、民居皆沿溪、河而造。乌镇与众不同的是沿河的民居有一部分延伸至河面，下面用木桩或石柱打在河床中，上架横梁，搁上木板，人称"水阁"，这是乌镇所特有的风貌。水阁是真正的"枕河"，三面有窗，凭窗可观市河风光。水阁是乌镇的独创，是乌镇的魅力所在。

百步一桥。桥是江南水乡古镇不可或缺的因素。乌镇历史上桥梁最多时有120多座，真正是"百步一桥"，现存30多座。其中西栅有通济桥、仁济桥，中市及东栅有应家桥、太平桥、仁寿桥、永安桥、逢源双桥，南栅有福兴桥和浮澜桥，北栅有梯云桥和利济桥，具有浓厚的历史文化气息。

Major Commercial Activities 主要商业业态

Wuzhen business district is centered by Zhongshi. The four areas, east area, west area, south area and north area, have their own system and focus. There is a perfect and complete business net. The east area is centered by "Caishen Bay", the town's largest distribution center for aquatic products. There are kinds of famous specialties and typical commercial buildings to create a strong business climate, which is different from other water towns.

　　乌镇商业区以中市为中心，东南西北四栅各成系统，又有各自的中心，形成了较为完善的商业网络。东栅的中心是财神湾，全镇最大的水产集散中心、批发零售商徐恒裕东号和西号都在这里。在乌镇，随处可以买到著名特产，如姑嫂饼、杭白菊、三珍斋酱鸭、乌镇羊肉、熏豆茶、三白酒等，其中尤以"姑嫂饼""杭白菊"最为有名。另外访庐阁茶馆、高公生糟坊、宏源泰染坊、汇源典当行等商业建筑也在常丰街，具有浓郁的商业氛围，这与其他江南水乡古镇有很大不同。此外，还有锦和斋经营南北货物，声誉极盛，至今尚存。

Hongyuantai Dyehouse

Founded in Song and Yuan dynasties in south area, the dyehouse was moved here since Guangxu Period of Qing Dynasty. Blue calico products are produced and distributed here.

宏源泰染坊

宏源泰染坊始创于宋元时期，原址在南栅，清光绪时期迁址于此，是蓝印花布的制作基地，也是蓝印花布制品集散中心。

Shunchanghao Fur Shop

The fur shop is a good stage for visitors both at home and abroad to learn about the influence and history of fur in Tongxiang. Modern fur in an old shop not only becomes a useful supplement for Wuzhen culture, but also brings visitors warm in the cold winter.

顺昌号皮草行

顺昌号皮草行为广大海内外游客提供了皮草展示的平台，向世界展示桐乡在皮草行业深厚的实力和历史渊源。老店中的新品，不仅成了对乌镇文化的有益补充，也是古镇在寒冷的冬天带给游客的一份温暖。

Fate Bar

It is a fate-themed bar, decorated with many Buddha statues from the inside to the outside, which implies that you can keep calm in your heart even though you are in a bar. The bar has a very strong nostalgic atmosphere with enchanting frescoes and ambiguous red curtain lights.

五百回酒吧

　　"前世的五百次回眸，才换来今生的一次擦肩而过。"五百回酒吧是一个以缘分为主题的酒吧，从里到外摆放着多尊佛像，完美诠释了"酒肉穿肠过，佛在心中留"的生活意念，酒吧设计具有非常浓厚的怀旧氛围。

Xuchang Sauce and Pickle Shop

The shop has a long history of one hundred years. Founded in 1859, it's the first sauce shop in Wuzhen. With traditional hand brewing method, its products are delicious.

叙昌酱园

叙昌酱园是乌镇的百年老店。清咸丰九年（1859年），乌镇人陶叙昌创立叙昌酱园，这是乌镇有史以来最早的酱园。创立之初，主要经营豆瓣酱、酱油、酱菜，所产酱品行销嘉兴、太湖等地区。叙昌酱园的酱品采用传统手工酿制法，每年春秋，酱园收购邻近村镇的优质黄豆、蚕豆、小麦等原料，利用竹匾制曲，经过长达半年的自然晒露、发酵而制成。

Yida Silk Shop

It's a big folk workshop related to every process of silk production, and a microcosm of long silkworm culture. The hall of the silk shop has a good name with the meaning of good luck and everything going well. The hall is more spacious and luxurious than normal houses. There are fish and flower patterns on the ceiling, which means wealth.

益大丝号

益大丝号是一家集养蚕、种桑、收茧、缫丝、制丝、造锦于一体的民间大作坊，这里展示的是整个杭嘉湖平原悠久的蚕文化缩影。益大丝号的正厅称万和堂，取万事和谐之意，大厅的布置比一般民宅宽敞豪华，天棚板上雕有鱼跃龙门、四季花卉的图案，象征财源广进，店铺出售本家生产的乌锦、丝绵等物品。

Zhang Hengxing Lantern

A giant horse lantern at the entrance attracts the attention of passers-by. There is an array of lanterns hanging in the whole hall, and there are nearly 20 varieties and more than one hundred specifications, dazzling and dizzying.

张恒兴灯笼铺

张恒兴灯笼铺门口一盏巨型走马灯吸引过往行人的眼球。进入店铺，形态各异、琳琅满目的花灯在店堂里悬挂得满满当当，有球形灯、直筒灯、五角灯、蛋形灯、宝塔灯、异型灯、卡通灯、镭射灯等近20个品种、百余种规格，让人眼花缭乱、目不暇接。

Lizhi Academy 立志书院

Lizhi Academy is located at the former residence of Mao Dun, and founded in 1865. In 1904, it was changed to be Lizhi Primary School, and became the first primary school in Wuzhen. Mao Dun was one of the first students.

立志书院位于茅盾故居东侧，清同治四年（1865年）创办，清光绪三十年（1904年）改名为立志小学，是乌镇第一所初级小学，茅盾为第一批学生之一。

Qinggeng Yudu 晴耕雨读

It's an academy of classical learning, embodying Chinese farming community mainstream value.

"男耕读女纺织家承百年之业，藏诗书教子孙文起八代之衰。""晴耕雨读"体现的是中国几千年农耕社会的主流价值观，是"耕读传家"形象说法。

Wenchang Pavilion 文昌阁

It's located on the river port in the front of Lizhi Academy. There is only a narrow street between them. In old times, scholars accompanied by servants came here by boat. After imperial civil examination was abolished at the end of Qing Dynasty, it became a place for visitors to relax. Meanwhile, for the long-time center location, it is also the town's media center.

文昌阁位于立志书院门前河埠上，书院与阁之间仅隔一条不宽的观前街。旧时读书人到文昌阁，一般都是乘坐小船前来。清末科举废止，文昌阁便成了人们游玩的地方，同时，由于长期以来造就的中心地位，它又是镇上人的新闻传播中心。

Wu Fire Hose Station 乌青水龙会

It functions as present fire station and plays an important role in the history of Wuzhen. The main building material is wood. It features wood grids, which not only promote ventilation, but also attract visitors.

据镇志记载，每年的农历五月二十日是分龙日。这天下午，各坊水龙聚合在西寺白场上演练，谓之"演龙"，各坊水龙"齐集旷地，竞演水力，各显技艺"，热闹壮观，实为一次精彩的消防演习。

Zhaoming Academy 昭明书院

The academy sits north towards south. It's a two-level ancient building complex with gable roof and semi-ambulatory.Its main building is a liberary.

昭明书院坐北朝南，是半回廊二层硬山式古建筑群。现在主楼为图书馆，收藏有文化、社会科学、艺术、休闲旅游等方面的图书和杂志，并设有电子阅览室、讲堂、书画、教室等。

Nanzhao Ancient Street, Dali
大理南诏古街

History 历史承袭

According to historical records, Weishan Ancient City in Tang and Song dynasties was a religious venue and was guarded by the Duan family in Yuan Dynasty. After the emperor Zhu Yuanzhang established the Ming Dynasty, he wanted to consolidate the central government to Yunnan Province, so Menghua, current Weishan, a place of strategic significance, was chosen as the site to build a defensive city. In 1390, a city of four gates was built, with moat and roads around the wall, and drawbridges installed outside the gates. The north tower of three-storey-high has a Moon City outside. The Xinggong Tower at the city center links to the east, west, south and north streets.

In 1938, the ancient city has been rebuilt and expanded in order to adapt to the needs of economical and social development. The Moon City was rebuilt into a small park, and the north wall was pulled down for the current east and west streets. The Gongchen Tower was expanded and more than 50 shops were built and priced by the government to sell to the public. Finally, the Sifang Square took in shape. The layout of the ancient city has also changed by that time, with the original center replaced by Gongchen Tower, 390 m to the north, and linked to the Risheng Street, Yuehua Street, etc., formed a central axis. The Nanzhao Ancient Street is just composed of the well-preserved Yuehua Street, Risheng Street and South Street of gorgeous history.

据史籍记载，巍山古城在唐宋时为宗教场所，元朝段氏土总管筑土城据守，朱元璋建立明朝后，为稳固经营云南，遂选择了在历史上具有重要战略地位的蒙化（今巍山）筑城。明洪武二十三年（1390年）建起的古城，周长2千米，城墙高2丈、厚2丈，砖石城墙，有垛头1 207个，垛眼430个，有4座城门，门上建楼，东曰忠武、南曰迎薰、西曰威远、北曰拱辰，城方如印。城墙外四周有护城河、驰道，城门外设吊桥，北门城楼有三层，外建月城，城中心处建星拱楼为印柄，向四面延伸建东街、南街、西街、北街。

"民国"二十七年（1938年），时任蒙化县长的宋嘉晋为适应经济和社会发展的需要，对古城进行改建、扩建，拆除月城建小公园，拆除北城墙建新东街、新西街，拓宽拱辰楼四周，由政府统一建盖铺面50余间，标价向社会出售。由此，围绕拱辰楼形成了一条新的商业街——四方街。古城格局发生了变化，原古城中心星拱楼变为次中心，拱辰楼则成为古城中心，城市中心向北移390米，与左氏土知府、日昇街、月华街连成一片，形成了南门城楼、星拱楼、拱辰楼、文献楼（现群力门）一条中轴线，"南诏古街"就由保留至今的月华街、日昇街、北街和南街组成。

Location 区位特征

Located at Dali, Yunnan Province, Nanzhao Ancient Street is mainly composed of traditional architectural structures which retain the plain and elegant layout as well as traditional businesses of local place, and is called a "live ancient street". Due to its strategic geographical position as well as prosperous business, it has always been the political, economical and cultural center of western Yunnan. It has nurtured a large group of people adapting to all kinds of taste as well as a diversified food culture.

南诏古街位于云南省大理白族自治州巍山彝族回族自治县，建筑多为大理传统的土木结构，保留了古朴典雅的布局和传统商业，是一条"活着的古街"。因地处要冲，南来北往、西进东出的商贾云集，他们南腔北调，人气旺盛，故这里一直是滇西的政治、经济、文化中心，形成了饮食文化中能适应酸、辣、甜、咸、淡各种口味的人群，造就了饮食文化的多元和丰富。

Market Positioning 市场定位

Nanzhao Ancient Street is positioned: relying on local history and culture, to build a tourism complex with Nanzhao Palace as the brand, to develop a tourism and culture center of tourism, leisure, holiday, entertainment, business and property, and to realize the harmonious combination of culture, business and tourism. Moreover, industrial structures should be optimized; historic remains should be protected; and tourism industry should be upgraded, so as to build the site into a featured scenic spot for local culture exhibition as well as an important carrier for local tourism transformation and upgrading.

以南诏历史文化为主线，打造以"南诏王宫"为品牌的旅游综合体，形成集观光、休闲、度假、娱乐、商业、置业等多功能于一体的旅游文化产业聚集区，实现文化、商业、旅游的高度契合，促进产业结构优化、古城遗迹保护和旅游产业升级，成为展示巍山地域文化的特色景区、巍山县旅游业转型升级的重要载体。

Street Planning 街区规划

The program sets a development guideline of "relying on Nanzhao culture to protect and develop the city, streets, historic relics, temples and music." A series of protective development activities are taken according to local conservation regulations to plan the street.

First and foremost, all original roads are renovated as black stone roads, with all exposed lines and pipes hidden underground. Two stone memorial archways along the roads are also restored, yet road lamps, greening, and sanitation facilities are completed. The whole reconstruction work can be divided into four phases.

设计师制定了"以南诏文化为主线，保护开发古城、古街、古遗址、古庙、古乐"的发展思路，按照《巍山县历史文化名城保护条例》，对古城进行了一系列保护性开发。

将原道路恢复为青石板路面，对外露的供电、通信、有线电视、给排水管线进行地下隐藏，恢复重建两座石牌坊景观，完善路灯、绿化、环卫等配套设施，先后分四期完成了北街、日升街、月华街、南街路面改造及临街建筑风格整治。

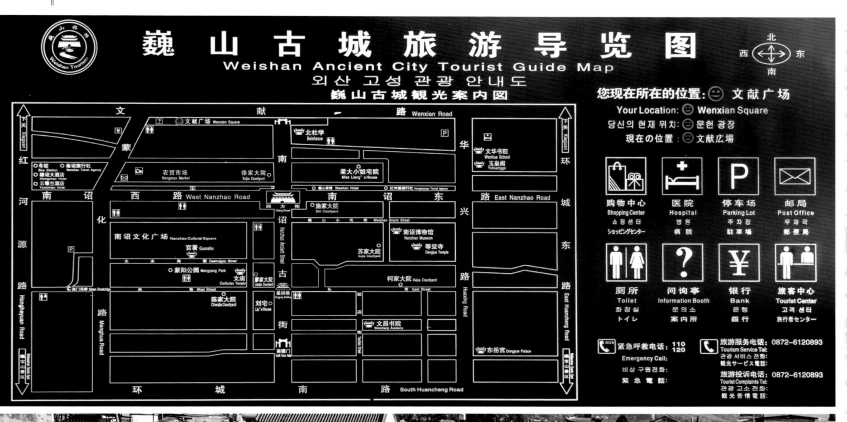

Street Design Features 街区设计特色

Most folk houses in this district have retained the original features of Ming and Qing dynasties. The ancient street connects four old towers and forms harmonious landscape.

There are only one or two shops along the street interface of Risheng and Yuehua Streets, while the back is bamboo-type folk houses of three entrances and two courtyards. All buildings are simple and elegant, with carved beams and painted rafters. Since local residents pay much attention to façades, most buildings are designed with single or double eaves, wooden structure roofs and overhanging brackets. Each household features courtyards as well as orchid flowers.

The numerous buildings along the street can be dated back to Qing and Ming dynasties. They have traditional historical features of the site largely retained and fully interpreted the mean of the word "ancient". They are soul of the ancient street.

民居大多保留了明清建筑风貌，多为土木结构。青瓦坡顶、"三坊一照壁"、"四合五天井"，古街串联四座古楼，头尾衔接，相得益彰。

日昇、月华两街多为一间或两间铺面，后面是三进两院的竹筒式民居。建筑古朴典雅，雕梁画栋。民居讲究门面，大门多为单檐或重檐，木架瓦顶，出阁架斗，户户有庭院，家家栽兰花，是"满城春兰风亦香"的真实写照。

古街上汇集了古楼、古坊、古民居、名优特产、风土人情，古街两旁有玉皇阁、文华书院、文昌书院等众多明清建筑群，较完整地保留了传统历史风貌，诠释了"古"字的全部含义，构成了历史文化名城跳动的灵魂。

Major Commercial Activities 主要商业业态

There are three theme streets covering different commercial activities of the site. The ancient folk houses and archaic courtyards along the streets are saturated with the profound cultural deposits of Weishan Ancient City.

There are many traditional cuisines as well as local snacks at the banquet on the streets. Each ethnic group features unique flavors and food preference, and all those dishes are delicious.

　　街区包括古玩、珠宝、玉器一条街，娱乐休闲吧、历史文化表演、民族风情一条街，文化休闲、琴棋书画馆、笔墨纸砚馆、茶馆、咖啡馆、曲艺馆一条街，土特产一条街，民族服饰一条街，工艺小商品一条街以及现代商品一条街。名店有"重兴店""合义老店""兴隆店""茶升店""百宝生""唐记老客栈""李记世生客栈"等。两旁的古民居、古色古香的庭院，浸润着巍山历史文化名城丰厚的文化底蕴。

　　在饮食方面，这里有传统的"南诏宴席""八大碗""两滴半""三滴水"还有小吃"肉饵丝""牛打滚""一根面""糖火烧"、咸菜、蜜食等。彝、回、汉、白、苗等民族各有其独特的饮食风味喜好，如彝族喜爱的苦荞粑粑、米血肠、鸡血饭、坨坨肉，都美味无比。

Leisure 休闲类

Qi Huang Ge

Architectural structures such as lattice door and davit are partly decorated with wood carvings in shape of rolled grass, flying dragon, bat, white rabbit, etc. All those patterns, regardless of animals or plants, are ever-changing and are applied skillfully. The multilayered landscape, figures, animals and plants on the buildings are vivid.

麒煌阁

　　建筑的格子门、横披、板裙、耍头、吊柱等部分用木雕装饰。卷草、飞龙、蝙蝠、玉兔，各种动植物图案造型千变万化，运用自如，如"金狮吊绣球""麒麟望芭蕉""丹凤含珠""秋菊太平"等情趣盎然的图案作品。把多层次的山水人物、花鸟虫鱼表现得栩栩如生。

Mengshe Inn

Inn courtyard is the most typical architectural structure in Bai nationality. There are front and back yards in the inn. The interiors are filled with simple and warm homey feelings.

蒙舍驿站

客栈庭院为典型的白族建筑结构，分前后两院，内部设计为家庭式温馨简约风格。

Mo Shang Hua Kai

The main courtyard of Mo Shang Hua Kai features six doors engraved with flower and bird patterns. In the wing rooms on two sides, there is a wooden lattice window of colorful patterns. The floor paved with bricks, pebbles, hexagonal and octagonal bricks and tiles seems natural and plain.

陌上花开

正宅安装有六扇雕刻着花鸟的合门，图案有"喜鹊登枝"、"丹凤朝阳"、"鸳鸯戏水"、"仙鹤青松"等。两侧的厢房分别有一扇花格子木窗，窗上的图案丰富多彩，有梅花窗、古钱窗、绣球窗……地上则有用方砖、卵石、六角砖、八角砖和瓦片铺设而成，显得自然古朴。

Shang Ceng Mo Fang

Arched gate of rich decorations is a general performance of the architectural patterns in Bai nationality. And that is just what Shang Ceng Mo Fang looks like, where palace form is adopted, and clay sculptures, wooden carvings, colored paintings, stone inscriptions, Dali stone screens, and embossed blue bricks form rich and colorful 3D patterns splendid yet graceful.

尚层魔方

富于装饰的门楼是白族建筑图案的一个综合表现。采用殿阁造型，再以泥塑、木雕、彩画、石刻、大理石屏、凸花青砖等组合成丰富多彩的立体图案，显得富丽堂皇又不失古朴大方。

Yilouyoumeng Bar

In Yilouyoumeng Bar, both brick column and the tiling are rendered at the joints, while wall body is painted white. Colored paintings at the cornices are varied in widths and adorned with decorative ribbons in different colors. As a result, the whole bar looks refreshing and elegant.

一楼幽梦酒吧

墙体的砖柱和贴砖都刷灰勾缝，墙心粉白，檐口彩画宽窄不同，饰有色彩相间的装饰带，以各种布置表现出一种清新雅致的情趣。

引导指示系统

Guidance & Sign System

Dayan Old Town, Lijiang
丽江大研古城

History 历史承袭

Lijiang Old Town was once the governmental office location during Ming and Qing dynasties. After the Republic of China, it was renamed as Dayan Town.

The ancient town was initially built in Song Dynasty and so far it has been stood here for more than 800 years. It is located on the vital communication road of Yunnan, Sichuan and Tibet, and those high mountains at surroundings are its natural hedge. With flourishing business as well as developed transportation, the old town became an important trade hub since its foundation. Traveling merchants from the southwest border of China gathered here for exchanging local special products and daily suppliers. For once a time, the town was a very important transportation hub between China and India.

丽江古城，又名"大研镇"，曾是明朝丽江军民府和清朝丽江府的府衙署所在地，明朝时称大研厢，清朝时称大研里，民国以后改称大研镇。

古城始建于宋朝，丽江木氏先祖将统治中心由白沙迁至现狮子山，至今已有八百多年的历史。古城地处滇、川、藏交通要道，在土府时代不筑围墙，以四周的高山作为天然屏障。自建成以来，商旅云集，各路马帮往来不断，大研古城成为重要的贸易中转站。木里、源盐、永宁、下关、大理、维西、中甸、拉萨等地的客商汇集于此，交换各种土特产品及日用品，曾一度成为当时中国通往印度的重要集镇。

Location 区位特征

Dayan Old Town is located at the foot of Jade Dragon Snow Mountain, the center of Lijiang Dam. It leans against several mountains in the north and overlooks tens of miles of fertile lands and wild fields to the southeast. Standing 2,400 meters above the sea level, the town is home to former local administrative organization and a national historic and cultural city. It is praised as the "Oriental Venice" as well as the "Suzhou on Plateau" for its beautiful water town scenery, unique layout and architectural style.

丽江大研古城坐落于玉龙雪山下，处于丽江坝中部，北依象山、金虹山，西枕狮子山，东南面临数十里良田阔野。古城海拔2 400米，是丽江行政公署和丽江纳西族自治县所在地，为国家历史文化名城，以江南水乡般的美景、别具风貌的布局及建筑风格特色，被誉为"东方威尼斯"和"高原姑苏"。

Market Positioning 市场定位

Dayan Old Town takes "natural, effective, sincere, inclusive" as development and construction concept to build a well-known historic and cultural city of flowing city space, vibrant water system, unified architectural complexes, and pleasant living environment.

项目以"崇自然、求实效、尚率直、善兼容"为开发建设理念，打造具有流动的城市空间、充满生命力的水系、风格统一的建筑群体、亲切宜人的空间环境的中国历史文化名城。

Street Planning 街区规划

The layout of the old city makes use of the most of existing topography as well as surrounding natural environment to protect the city from chilly wind from the northwest and bring in natural daylighting from the southeast. The Yuquan River originated from the foot of the Elephant Hill at the northern town diverges into three tributaries after it enters into the town and then divides into more branches running throughout the ancient land, forming riverbank scenery at each household.

Centered on the Square Street, all streets are freely spread over the town, with Xinhua Street, Wuyi Street, Qiyi Street, Xinyi Street and Guangyi Street served as main axes. All streets and lanes are arranged along the water. Over 300 old stone bridges over rivers, trees, old lanes and houses form aesthetic water town scenery on a plateau. Natural springs are fully used to build wells for daily water, a wise wisdom passed on from Naxi ancestors, which reflects the harmonious relationship of man and nature. Spaces, closed or open, form a well-developed city road network accessible to all corners. The Mu's Mansion in the town is extraordinary grand.

As for street planning, layout, structure and architectural form are all arranged according to local specific conditions and traditional living habits, and are combined with the traditions in folk houses of Han, Bai and Tibetan nationalities. All buildings have taken bold innovation in terms of anti-seismic capacity, shading, rain-proof, ventilation and decoration to form unique style. The most brilliant is its free organism showing natural and plain creativity, which is a rare important heritage for studying the Chinese architectural history and culture.

　　古城在布局上充分利用山川地形及周围自然环境，北依象山、金虹山，西枕猴子山，东面和南面与开阔的坪坝自然相连，既避开了西北寒风，又朝向东南光源，形成"坐靠西北，放眼东南"的整体格局。发源于城北象山脚下的玉泉河水分三股入城后，又分成无数支流，穿街绕巷，流布全城，形成了"家家门前绕水流，户户屋后垂杨柳"的景致。

　　街道以四方街为中心，不拘于工整而自由分布，形成以新华街、五一街、七一街、新义街、光义街5条街道为经络的格局。主街傍水，小巷临渠，300多座古石桥与河水、绿树、古巷、古屋相依相映，极具高原水乡"古树、小桥、流水、人家"的美学意韵。充分利用城内涌泉修建的多座"三眼井"，上池饮用，中塘洗菜，下流漂衣，是纳西族先民智慧的象征，充分体现人类与自然的和谐统一。空间时而封闭，时而开阔，组成了一个通达全城的路网，城中的木氏土司衙署则呈现出一派"宫室之丽，拟于王者"的非凡景象。

　　规划在布局、结构和造型方面按自身的具体条件和传统生活习惯，结合了汉族以及白族、藏族民居的传统，并在房屋抗震、遮阳、防雨、通风、装饰等方面进行了大胆创新，形成了独特的风格。最鲜明之处就在于无统一的构成机体，显示出"依山傍水、穷中出智、拙中藏巧、自然质朴"的创造性，是研究中国建筑史和文化史不可多得的重要遗产。

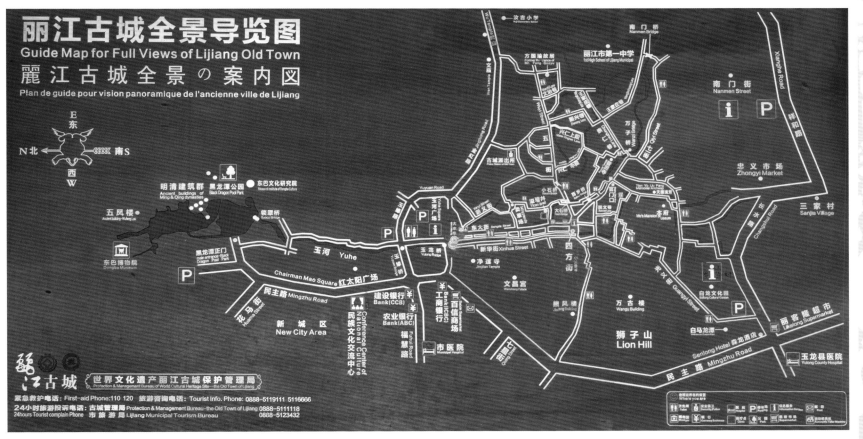

Street Design Features 街区设计特色

There are rows upon rows of Naxi style folk houses in the old town. The exquisite arch bridges in front of each building span across crystal streams running over the town, while willow branches beside water sway gently in the breeze. Black stone lanes zigzag throughout the town and connect with each other. As a result, unique water town scenery on plateau is created.

Ancient Street

All streets are paved along the mountains and waters with red breccias. The paving of delicate texture is clean and the patterns are natural and elegant. It is harmonious with the whole environment.

Bridge

There are 354 bridges over the Yuhe water in the town. There are gallery bridges, stone arch bridges, slab bridge and plank bridges, etc. Many famous bridges here are remains of Qing Dynasty. Dashi Bridge, the leader of the bridges in the town, is located 100 meters to the east of the Square Street, and was built by Chieftain Mu. It is also called "Yingxue Bridge", because one can see the reflection of the Jade Dragon Snow Mountain over the water under it. It is a double-span stone arch bridge and the deck is also paved with traditional breccias. Its gentle arch is very convenient.

Folk Houses

The folk houses in the town are epitome of Naxi architectural art and style. Most façades are paved with stone plinths and walls are rendered with plasters. There are also tiling in the wall corners as well as black tile roofs. The color is harmonious and the outer silhouette is beautiful. Based on the original wooden houses in Naxi nationality, the design mingles the architectural features in Han, Bai and Tibetan nationalities. Therefore, both architectural layout and arts are of distinctive local features and ethnic styles.

Generally, most buildings in the town are two-storey wooden structures about 7.5-m-high, with tenon-and-mortise structures, mud retaining walls, tile roofs, and verandas. According to different frameworks and verandas, they can be divided into four different types of houses. The most typical two houses are households with three squares and one screen, as well as the households with an enclosed courtyard and two patios.

The framework is very flexible. There are various joints, and the whole structure is oblique inward for one degree to increase general stability. At the point of junction, various tenons are used according to specific force. The parapet wall with heavy bottom and light top is solid and practical; it is helpful to resist earthquake.

Courtyards in local folk houses are paved with cobblestones and red breccias. All patterns are designed according to the size of the courtyards and the preference of house owners, with content rich in flowers, animals, the Eight Trigrams, folk legends and fairy tales, design techniques of primitive simplicity, and layout orderly. Mansions with large size and multiple courtyards are generally composed of two squares and one screen, flower bed, pool and etc.

古城纳西风格浓郁的民居鳞次栉比，家家门前精巧的拱桥横跨穿城过巷的清澈溪水，排排垂柳在轻风中摇曳，青石板铺就的小巷蜿蜒曲折、纵横交错，形成了"家家泉水，户户垂杨"与"小桥、流水、人家"独特的高原水乡风貌。

古街

街道依山势而建，顺水流而设，质感细腻，以红色角砾岩（五花石）铺就，雨季不泥泞、旱季不飞灰，石上花纹图案自然雅致，与整个城市环境相得益彰。

古桥

在古城区内的玉河水系上架有354座桥梁，其密度为平均每平方千米93座。形式有廊桥（风雨桥）、石拱桥、石板桥、木板桥等。较著名的有锁翠桥、大石桥、万千桥、南门桥、马鞍桥、仁寿桥，均建于明清时期。大石桥为古城众桥之首，位于四方街东向100米，由明代木氏土司所建，因从桥下河水可看到玉龙雪山倒影，又名"映雪桥"。该桥系双孔石拱桥，拱圈用板岩石支砌，桥长10余米，桥宽近4米，桥面用传统的五花石铺砌，坡度平缓，便于两岸往来。

民居

古城的民居是纳西族建筑艺术和建筑风格的集中体现。在体形组合及轮廓造型上纵横交错，外观立面多为石砌勒脚，墙面抹灰，墙角镶砖，青瓦铺顶，色调和谐，轮廓优美。设计在吸取纳西族原始的"井干式木楞房"的基础上、融汇了汉、白、藏等民族建筑的特点，在布局形式、建筑艺术等方面具有鲜明的地方特色与民族风格。

建筑一般是高约7.5米的两层木结构楼房，为穿斗式构架、垒土坯墙、瓦屋顶，设有外廊（即度子）。根据构架形式及外廊的不同，可分为平房、明楼、雨步厦、骑度楼、蛮楼、闷楼、雨面厦等七大类。布局形式有"三坊一照壁""四合五天井""前后院""一进两院""两坊拐角""四合院""多进套院""多院组合"等类型，其中，以"三坊一照壁"和"四合五天井"为典型。

古城民居的构架处理十分灵活，设有"勒马挂""地脚""穿枋""千斤"等具有拉结作用的构件，整个构架按百分之一的斜度使柱头往里倾斜、柱根部向外展开，增强了构架的稳定性。在构架的连接节点部位，根据受力情况，分别使用"两磴榫""大头榫""平插榫"，井设暗梢等柔性节点，"下重上轻"的护墙体坚固实用，以利于抗震。

民居的庭院则采用鹅卵石、五花石等铺装，图案根据庭院大小或房主喜好而定，内容涉及花鸟鱼虫、八卦阴阳、民间传说、神话故事等，手法古朴，布局严谨。占地大、院落多的宅院，普遍由两坊一照壁、花台、水池等构成。

Major Commercial Activities 主要商业业态

Major commercial activities in the town can be divided into four categories, namely accommodation, dining, shopping and recreation.

All four commercial activities are balanced in scale. Generally, accommodation which occupies 37% ranks at the first, while shopping and dining, 25% and 23%, follow suit. The last is the recreation 15%.

古城主要业态大致可以分为四类，即住宿(特色客栈)、餐饮(主题餐馆、特色餐厅)、购物(特色购物商店)、休闲娱乐(酒吧、咖啡馆、书吧等)。

从业态配比上来看，四类业态相对均衡，没有占绝对性的业态类型。总体来看，住宿所占比重较大，达到了37%；其次是购物和餐饮，分别占25%和23%；最后是各种休闲娱乐类场所占15%。

Food & Beverage 餐饮类

Zui Ba Xiang Restaurant

Zui Ba Xiang Restaurant is a local specialty restaurant that located on No.107 Zhongyi Lane, Guangyi Street.

嘴巴香菜馆

嘴巴香菜馆位于古城光义街忠义巷107号，特色菜有纳西烤鱼、杂锅菜、素锅、藏民家烤牦牛肉、烤乳猪、山药鸡汤、牦牛干巴、纳西烤肉、鸡豆凉粉等。

Accommodation 住宿类

Tree's Tale Hostel

Tree's Tale Hostel is located near the back door of Mu's Mansion, within walking distance to the Square Street and Wangu Building. The restaurant launched on April 1, 2010 is a two-storey building with 8 rooms. It enjoys elegant environment as well as convenient traffic.

一棵树客栈

一棵树客栈地处古城木府后门的过街楼附近，四方街、万古楼近在咫尺，于2010年4月1日开业，楼高2层，客房总数8间（套）。环境优雅，出行便利。

Da Yan Garden

Da Yan Garden is a Naxi style courtyard hotel of vivid carvings on the doors and windows.

大研阁酒店

大研阁酒店为纳西族庭院式酒店，高原水乡中的客栈在院中五彩缤纷的鲜花的映衬下春意盎然。小楼的门窗精雕细刻，花鸟图案相映成趣，栩栩如生。

Chi He Jing Pin Hui Suo Inn

Covering 1,600 square meters, Chi He Jing Pin Hui Suo Inn is close to tourism attractions such as Baimalongtan Temple and Mu's Mansion. It takes only 5 to 8 minutes to the Square Street and 3 minutes to the parking lot as well as bus station at Bailong Cultural Garden. There are leisure zone, tea room, swings, book bar, view platform, vegetable farmlands and many other facilities.

驰和精品会所客栈

此客栈占地面积达1 600平方米，邻近白马龙潭寺、木府等景点，步行5～8分钟即可至四方街，步行3分钟即可出古城到达白龙广场停车场、公交站，周边风景秀丽，出行便利。客栈内有休闲区、品茶吧、秋千、书吧、自助厨房、烧烤平台、机麻室、观景台、菜地等多类配套设施。

Spring Inn

Spring Inn of 800 m² only has 18 guest rooms. With more than 100 years of history, the Naxi style building offers guests tea room, study, western restaurant, cafeteria and many other facilities.

叠泉酒店

叠泉酒店建筑面积为800平方米，仅18间客房，建筑有百年历史，为纳西族传统的"三坊一照壁"布局，酒店提供茶室、书房、西餐厅、自助厨房等配套设施。

Courier Inn

Courier Inn is the only inn in Lijiang with Naxi folk house appearance and star level interior décor. There are totally 60 rooms and suites, and the inn connects directly to the bars in the town as well as the Square Street. It has retained the architectural style in Ming and Qing dynasties, and from here one can enjoy a good view of the whole city.

桃花岛驿站

桃花岛驿站是丽江古城内唯一具有纳西民居外观、房间内饰按星级标准配置、真正具有小桥流水环绕的古城客栈。配备各类房型共60间（套），包括复式家庭套房、观景房、双床间等。驿站与古城酒吧一条街、四方街逐一相连，保持了明清时期的建筑风格，青砖古瓦，瑞阁楼绣房，可一览古城全貌。

Lost in Venice Inn

Lost in Venice Inn is a courtyard of typical Naxi style. There are three-storey wooden houses which are rare in the town. The inn is built beside water, with exquisite courtyards as well as tranquil and elegant environment. Nearby famous attractions include Mu's Mansion, Dashi Bridge, Xiaoshi Bridge and so on.

威尼斯迷路客栈

此客栈为典型的纳西院落，客房主楼是古城少有的三层木房，临水而建，精致的院子，清幽雅致。客栈紧邻木府、大石桥、小石桥、三眼井等著名景点，万古楼、纳西古乐会、黑龙潭公园、丽水金沙等景点近在咫尺。

Lijiang Old Town Castle Hotel

Lijiang Old Town Castle Hotel is initially founded in Qing Dynasty by master Miaoming based on Naxi architectural style. The hotel is named for its special position opposite to the Jade Dragon Snow Mountain, which looks as if it were plunged into the mountain. There are numerous precious trees as well as pavilions, pagodas and towers spread over a tranquil environment. For more than 200 years, it has provided all informed scholars and social celebrities around the country with an ideal place for enjoying scenery and communication.

嵌雪楼

嵌雪楼酒店建于清嘉庆至道光年间，由著名诗僧妙明法师按纳西族建筑风格建造，因正对玉龙大雪山，如同嵌在万古白雪之中，故名曰"嵌雪楼"。院内名木古树、亭台楼阁众多，环境幽静。二百多年来，一直是八方名士登临浏览、吟诗作画、扶琴和曲的好去处，声名远扬。登楼远眺，玉龙冰川、文笔朝霞、玉河烟柳、乡间田园村落、古城瓦屋鳞次栉比，四围景色尽收眼底，令人神清气爽。

Crafts 工艺品类

Bu Nong Ling

Bu Nong Ling is located at the end of Dashi Bridge, Sifang Street. It is the only one in Lijiang, and the other three shops in the world are located at Katmandu, Calcutta and Athens.

Compared with other shops, the bells in Bu Nong Ling are unique. Over each bell there hangs a piece of wood chip painted with Lijiang landscape and well-matched poem. The bells are handcrafted or machine-made. Each bell will produce unique sound. There are various bells such as bronze bell, copper bell, auspicious bell and peace bell.

布农铃

　　位于四方街大石桥桥头的布农铃小店，在丽江古城只有一家，在全世界总共有四家店。除了丽江，尼泊尔的加德满都、印度的加尔各答和希腊的雅典各有一个分店。

　　布农铃的铃铛与丽江其他小店里的铃铛比起来，外形与众不同：每一个铃下，都挂了一片布农所绘的丽江风景木片，背面有配合画面的诗意文字；铃分为手工制作和机制两种类型，每一个铃的声音都不相同。这些铃还分为青铜铃、古铜铃、阴铃、阳铃、挂在家里的吉祥铃、挂在车上的平安铃等。

Wushuang Artistic Jewelry Shop

All accessories in the shop are made of silver or small jewelries of best quality. The designers come from Hongkong, and there are several branches in the town as well as in Shuhe Ancient Town. The famous "Tile Cat" series with ferocious or lovely facial expressions can be placed over the roof to protect the family from bad luck, and the tiny silver tile cats are vivid and lovely.

无双艺术首饰

　　无双艺术首饰均用纯银或加小宝石镶嵌打造而成，做工精良，设计师来自香港，在大研和束河有多家分店。著名的饰品有 "瓦猫"系列，"瓦猫"面目专张狰狞又可爱俏皮，压在屋顶上以驱鬼辟邪、消灾免祸，以此形象设计的银饰，惟妙惟肖、小巧可人。

Bai Lü Tang

Bai Lü Tang mainly sells silver accessories, dinner wares, various crafts and scabbards.

百绿堂首饰

百绿堂首饰主营银首饰、银餐具、各种工艺品以及藏族同胞喜爱的刀鞘等。

Boke Town, Chengdu
成都洛带博客小镇

街区背景与定位

Street Background & Market Positioning

History 历史承袭

Luodai Ancient Town lasted for thousands of years is rich in legends and historical relics. There are old streets lined by Qing Dynasty buildings, showing a one street & seven lanes layout. Four Hakka halls, a Hakka Museum and a Hakka Park are gathered here, turning the place into a real Hakka town. More than 90% local residents are Hakka who still speak Hakka dialects and follow Hakka customs. The town is honored as the "First Hakka Town in West China".

洛带古镇历史悠久，相传汉代即成街，名"万景街"；三国时蜀汉丞相诸葛亮兴市，更名为"万福街"；后因蜀汉后主刘阿斗的玉带落入镇旁八角井而更名为"落带"（后演变为"洛带"）。唐宋时隶属成都府灵泉县（今龙泉驿区），明代时改隶简州（今简阳），清代时曾更名为"甑子场"。1955年，洛带区为简阳第十四区，辖10个乡。1976年，洛带区所辖的10个公社（原10个乡）划归龙泉，同时撤区建镇至今。镇内传说众多、古迹遍地，老街以清代建筑风格为主，呈"一街七巷子"格局，广东、江西、湖广、川北四大客家会馆，客家博物馆和客家公园坐落其中，是名副其实的"客家名镇、会馆之乡"。镇内90%以上的居民为客家人，至今仍讲客家话，沿袭客家习俗，被誉为"中国西部客家第一镇"。

Location 区位特征

Boke Town is located at the east Chengdu, the core of Luodai Ancient Town. It borders the Longquan Mountain Ridge on the east and links a wetland park on the west.

博客小镇位于成都东部、洛带古镇核心区，东靠龙泉山脉，西临湿地公园。

Market Positioning 市场定位

It is a high-end international cultural tourism town which gathers museums, cultural commercial streets, art studios, enterprise halls and five-star hotels.

小镇定位为集博物馆聚落、文化商业街、艺术工作室、企业会馆群、五星级酒店于一体的国际高端人文旅游小镇。

After analyzing the site condition, the land is divided into two parts, the riverside commercial streets in the north and the courtyard commercial blocks in the south.

1)Architecture

Riverside commercial streets are linked by riverside pedestrian streets and two commercial inner streets. Three commercial squares enclose central anchor shops to maximize customer flow. Courtyard commercial blocks are divided into four commercial complexes by main streets running from south to north and east to west. North-south main street links the entire project and boasts cultural and commercial activities. Squares are only used for guide pedestrian flow distribution, while linear spaces play a leading role. Through architecture combination and irregular dense arrangement, the buildings retain the texture of the ancient town and combine traditional landscaping techniques, adding fun to the commercial streets. The profoundness of the street space is a premium value of the program.

2)Landscape

The program pays much attention to an effective use of the greening and public space and coordinates with surrounding architectural forms to make the greening system a soft skeleton which shows each zone's image features. Both ornamental value and practicability need to be emphasized, yet sustainable landscape structure is required to be created to make ecological environment go along with functional construction, creating a benign ecological and ideal commercial environment. Green courtyards are to be planned and combined with patios, pavilion, corridor, and so on, to form a green landscape system of clear partition and rich space levels.

3)Traffic

The main entry lies at the intersection of San'e Street in the north and the planning road in the east, forming a corner square. While north-south main commercial street, west-east main street and outer ring road service as the main traffic roads, riverside commercial inner streets and courtyard commercial blocks are only for man use. Fire fighting trucks can enter the inner town from the east, south and north entries, run all the way along the ring fire lane surrounding the blocks and finally stop at the ecological parking lots by the road, when necessary.

在分析地块条件的基础上，将地块分为两大区域：一是北侧靠近河道的滨河商业街区，二是南侧的院落式商业街区。

1) 建筑

滨河商业街区由沿河步行街和两条商业内街将建筑串联起来，三个商业广场空间围合中心区域的商业主力店，最大化地吸引商业人流；院落式商业街区由贯穿南北和东西的主街分成四个组团式商业体，南北向的主街将滨河商业区和院落式商业区进行串联，以街道空间为主、开放空间为辅，在游人行走和停留间完成文化商业活动。因此，街道以线性空间为主导，广场只起到人流集散的辅助功能。通过建筑组合和不规则的紧密排列，延续古镇肌理，并将传统造景手法"对景""障景""框景"等运用其中，结合曲折的纹理，增加商业街的趣味性，形成街道空间的深邃感，表现出"空间增值"的特色。

2) 景观

注重绿化与公共空间的有效使用，并通过与周边建筑形态的协调处理，使绿化系统成为体现各区域形象特征的柔性骨架。首先，在创造景观的同时，兼顾实用性，从人的活动需求、心理感受出发，将形式与内容完美融合。其次，创造可持续发展的景观结构，坚持生态环境建设和功能建设的同步，创造良好的生态环境和理想的商业环境。同时，规划形成组团式的绿化庭院，结合天井建筑、亭、廊等小品，形成动静分区、空间层次丰富的绿化景观体系。

3) 交通

主入口位于北侧三峨街与东侧规划道路交界处，形成入口转角广场；南北向和东西的商业主街、外环道路为人车混行，滨河商业内街和院落商业街为步行商业街；消防车可由东、南、北三个主要车行入口进入内部，沿街区外环布置环形消防车道，环形消防车道外侧布置生态停车位，紧急时满足消防车道要求。

总平面布置图 1:900

There are varied courtyards hidden within streets and lanes. The Tulouba, Sanheping and Mini Shanghai are courtyard complexes mingling Sichuan, Fujian, Shanghai, Jiangxi and Anhui architectural styles. Each building is built according to grand mansions in this ancient capital in China by legendary masters for historical China images. Every household is independent from each other, yet they all show luxurious decoration. Winding wide streets and narrow lanes outline the leisure life of the town and hide prosperity inside.

The original commercial courtyard is epitome of exquisite Chinese life. It utilizes culture to carry on the business science within the streets, lanes and courtyards.

整个街区三街九巷千百院，院院不同。土楼坝、三和坪、小上海均为院落式商业集群，融四川、福建、上海、山西、安徽五派建筑风格于一体。每栋建筑均为藏品级定制，以古都名宅为用材标准，由名匠古法精工打造，一砖一木，皆精湛厚重。一院一字号，户户皆流金。宽街窄巷的迂回曲折，勾勒出悠然的小镇生活，藏繁华于内，立名门家业。

独创的商院是中国精致生活的缩影，以文化传承街、巷、院里的中国商道，可谓是一镇藏东方，一院会天下。

Major Commercial Activities 主要商业业态

In this project, architectural form, ecology, business activity and culture are combined together to create a nationwide first-rate cultural town from all aspects. With Luodai's mature tourism brands and convenient traffic, the Boke Town focuses on introducing specialty, boutique shopping, food & beverage, theme inns, so as to create unique theme streets and enrich tourism business activities. Diversified cultures including Hakka culture and intangible culture are blended into commercial activities.

　　"形态、生态、业态、文态"四态一体，全方位打造全国一流的人文小镇。依靠洛带古镇成熟的旅游品牌及便利的交通设施，一方面着力引入特色美食、精品购物、餐饮休闲、主题客栈等旅游性商业业态，打造休闲酒吧街、特色餐饮街等多条独具特色的主题街巷，丰富旅游商业业态；另一方面将多元文化融入商业业态中，包括客家文化、非物质文化遗产文化等。

Operation Measure 运营措施

Boke Town and Luodai Ancient Town will be blended seamlessly and the unified operation mode will be implemented to add premium value to the cultural tourism commercial circle of the ancient town.

　　博客小镇与洛带古镇实行无缝对接，采取统一规划、统一招商、统一推广、统一管理的"四统一"运营模式，整体提升洛带古镇的文化旅游商圈附加值。

Fengwu Academy　凤梧书院

Fengwu Academy formerly called Xunyang Academy was founded in 1694. In 1734, it is rebuilt and in 1787 it is renamed Fengwu Academy. It is still in operating in the early Republic of China.

　　凤梧书院原名寻阳书院，清康熙三十三年(1694年)由知州黄肇绅建于州城西门内。清雍正十二年(1734年)知州陈齐庶奉命重修，乾隆五十二年(1787年)知府常德改名"凤梧"，民国初仍有学生在此就读。

以文物保护为重点改建的商业街区

Commercial Streets Reconstructed on the Basis of Historic Preservations

Culture relics is a carrier from which human beings know themselves and acquire information as well as a monument of history and symbol of national self-respect. Therefore, it gradually becomes an important tourist resource. With the recognition of sustainable development and historic culture, commercial streets that can efficiently protect and make use of culture relics have emerged.

On the basis of historic preservations, the development of commercial streets lays its emphasis on deep exploration of intrinsic value of culture relics to show their cultural connotation, and through commercial activity to inherit and develop their cultural connotation.

Buildings: according to appearance, quality and storey and so on, different protection modes are applied.

Restoration. For culture relic protection sites and traditional buildings with good quality and appearance. we adopt method of restoration.
Improvement. This mode is mainly used in building concentrated area to protect its layout and appearance and at the same time focally adjust its interiors to improve the quality of buildings.

Renovation. It means to remove and reconstruct some danger buildings and untraditional buildings.

Rectifications. This mode is for buildings of good quality which are unharmonious with general environment to reduce degree of damage to historic environment.

Landscape: taking mutual benefit as principal harmonious coexistence of modern landscape and historical relics is created.

The design of landscape in the street as far as possible shows the culture connotation of relics in specific time to make visitors feel the profound value of these relics, therefore, consciously takes the responsibility of protecting historic and cultural relics. There are so many abstract cultural landscapes, such as relative stories of old celebrities, hiding in cultural relics. Therefore, in planning design, abstract culture is attached to concrete relics or modern landscape.

Commercial activities: bring visitor a concept of "Chinese traditional culture museum" by influence and leading role of time-honored shops.

Time-honored shops, which have won a widespread acknowledgement through the ages, are treasures of traditional commercial streets. When developing, governments often implement free or less tax for time-honored shop of local features. They also attract more old and famous brands to open branches here.

The commercial streets reconstructed on the basis of historic preservations pay more attention to the relationship between historical relic and consumers, and create a new concept and new space full of energy to lead the new consumption mode of "experience economy age".

　　文物不仅是历史的纪念碑和民族自尊的象征，也是人类认识自身、获得知识与信息的载体，因而它日趋成为一种重要的旅游资源。随着人们对可持续发展以及历史文化的重新认识，文物保护逐渐成为国内外的一种共识，在此背景下，能有效保护和利用文物的商业街应运而生。

　　基于文物保护的商业街开发，重点在于深刻挖掘文物资源的内在价值，展现文物所代表的文化内涵，并通过商业活动使之得到传承和发展，变"被动保护"为"主动保护"，真正实现"保护性开发"。主要表现为以下三个方面。

建筑：根据风貌、质量、层数等多种因素，分别采取不同的保护模式

　　还原。对文物保护单位以及沿街传统建筑中质量和风貌较好的建筑物、建筑群，采取还原的方式，修古如古，保证其内外风貌具有原真性。

　　改善。主要指建筑较集中的区域，这些建筑具有一定的保存价值、格局基本完整，但质量较差、难以适应现代生活需求。采用成片改善方式，重点对内部加以改造，以提高建筑质量。

　　更新。一是指对街区内无保留价值的危旧建筑和对传统风貌破坏较大的非传统建筑进行拆除并重建；二是指对不属于保护范围并难以改善的传统民居采取整修的方法。

　　整饬。对质量较好、但风貌与整体环境不协调、因历史原因难以立即拆除的非传统建筑，采用外立面整饬、层数削减的方法，使其与传统风貌相协调，降低对历史环境的破坏程度。

景观：以相辅相成为原则，使现代景观与历史文物和谐共生

　　在街区景观的设计中，尽可能地展现文物的特定内涵，使游客感受到文物的价值，从而自觉地承担起保护历史文物的责任。在规划设计中，常以抽象的人文吸引物附着于具象的文物资源或现代景观上；或将具有传统文化特色的民俗文化与博物馆相结合，增加浓郁的民俗风情；或设计雕塑景观，将人们想象中的历史人物表现在雕塑的实体之上；或以建筑群落等实体为依托，定期举办朝圣节会等。

业态：发挥"老字号"的影响力和带动作用，给人以"中国传统文化博物馆"之感

　　"老字号"是传统商业街上的宝贵财富，有着很好的口碑和知名度。在开发中，常采用政策倾斜的方法，对具有地方特色的传统"老字号"实行免收或少收税，吸引其他地方的"老字号"在街区开设分店，以聚集而形成规模。

　　以文物保护为重点改建的商业街区，更加注重历史文物与大众消费之间的关系，以人性化、亲民化的文化表达方式将文物内涵传递给消费者，创造出富有活力的新理念和新空间，引领"体验经济时代"的新消费模式。

Confucius Temple, Nanjing
南京夫子庙

Street Background & Market Positioning

街区背景与定位

History　历史承袭

Built in Song Dynasty, Confucius Temple is an old building group in a grand scale worshipping Confucius, the famous thinker and educator in ancient China. Confucius Temple had become a prosperous place since the Six dynasties. The Wuyi Lane, Zhuque Street and Taoye Ferry and so on are dwellings of great families at that time. In Ming Dynasty it became the imperial examination room, thus many service industries, such as restaurants, teahouses, gathered around it.

Due to the change of history, the bustling scene of Shili Qinhuai had passed out. Since 1984, National Tourism Administration and people's government of Nanjing have carried focal developing and restoring on Qinhuai to recover its Jiangnan street and commercial look in the late Ming and early Qing dynasties, therefore the Qinhuai area became a famous tourist attraction in China again.

夫子庙始建于宋代，是一组规模宏大的古建筑群，是供奉和祭祀中国古代著名思想家、教育家孔子的庙宇。六朝时代，夫子庙已成繁华之景，乌衣巷、朱雀街、桃叶渡等处，都是当时大族居住之处。在明代，夫子庙作为国子监科举考场，考生云集，因此这里集中了许多服务行业，有酒楼、茶馆、小吃摊等。古典戏剧《桃花扇》里所描写的"梨花似雪草如烟，春在秦淮两岸边，一带妆楼临水盖，家家粉影照婵娟"，十分贴切地展现了当时秦淮河上的繁华景象。

由于历史的变迁，十里秦淮昔日的繁荣景象早已不复存在。1984年以来，国家旅游局和南京市人民政府对秦淮风光带进行了重点开发、复建和整修，恢复了明末清初的江南街市、商肆风貌，秦淮河再度成为中国著名的游览胜地。经过修复的秦淮河风光带，以夫子庙为中心，包括瞻园、白鹭洲、中华门以及从桃叶渡至镇淮桥一带的秦淮水上游船和沿河楼阁景观，集古迹、园林、画舫、市街、楼阁和民俗民风于一体。

Location 区位特征

Confucius Temple is located at the north bank of Qinhuai River. The whole building group occupies a large area from Yaojia Lane at the east to Sifu Lane at the west.

In the front of the temple is a pool — Panchi. On its south bank there is a stone screen wall which is the biggest in the country. On its north bank is the Juxing Pavilion and Sile Pavilion. At the central axis there stands Lingxing Gate, Dacheng Gate, Mingde Hall and other buildings. On its east is Kuixing Maison.

夫子庙位于秦淮河北岸，整个建筑群占地甚广，南临秦淮北岸，从文德桥到利涉桥，东起姚家巷，西至四福巷，北临建康路东段。

庙前的秦淮河为泮池，南岸的石砖墙为照壁，全长110米，高20米，是全国照壁之最。北岸庙前有聚星亭、思乐亭；中轴线上建有棂星门、大成门、大成殿、明德堂、尊经阁等建筑；庙东有魁星阁。

Market Positioning 市场定位

A complex of various cultures, including cooking culture, sightseeing culture, custom culture and lantern show culture, to provide place for the masses' culture activities.

"以秦淮八绝为特点的饮食文化、以大成殿庙为首的景点文化、以各种旅游产品为内涵的民俗文化、以秦淮花灯为依托的灯会文化"的多种文化综合体，为群众文化活动提供场所。

Street Planning 街区规划

Confucius Temple centralizing Dacheng Hall has a north-south axis from the screen wall to Wei Mountain and buildings on both sides are symmetric. The buildings form a functional structure of "an area, three axes, several nodes and a net".

An area: the core area including Dacheng Hall, Examination Hall and inner Qinhuai River landscape belt.

Three axes: axis of Dacheng Hall, axis of Examination Hall and axis of inner Qinhuai River.

Several nodes: landscape nodes including buildings at main entrances, pavilions, memorial archways and ancient bridges.

A net: several historical streets and lanes organize a historic culture protection network.

All the buildings in this area should be lower than 12 meters. The cornices of three-floor buildings should be lower than 10 meters except some peripheral areas that can be lower than 18 meters to match up with the width of the street. The proportion of building height and street width should be within 1 : 1.

　　街区整体规划特点是"庙附于学，和国学、府（州）县学连为一体"，形成"前庙后学"的布局。夫子庙以大成殿为中心，从照壁至卫山，南北形成一条中轴线，左右建筑对称排列，占地约26 300平方米。形成"一片、三轴、多节点、网络化"的功能结构。
　　一片：以大成殿、贡院、内秦淮河文化景观带为核心片。
　　三轴：指的是大成殿中轴线、贡院中轴线、内秦淮河中心线这三条文化展示轴。
　　多节点：多个景观节点包括街区主要出入口建筑、楼阁、牌坊、古桥等。
　　网络化：指由多条历史街巷串联组织成历史文化保护网络。
　　整个街区内建筑高度控制在12米以下，大部分区域以1～3层建筑为主，3层建筑檐口高度控制在10米以下，外围部分区域结合街巷宽度，放宽至18米，建筑高度与街巷宽度比控制在1:1以内。

Street Design Features 街区设计特色

The buildings at Confucius Temple Qinhuai Scenic Area are of Ming and Qing dynasties style, and are gathering Dacheng Hall, Xuegong and Jiangnan Examination Hall which are the essence of Qinhuai scene. The Examination Hall Street beside the river is an antique tourist culture street of layout in ancient times.

The Dacheng Hall is located in front of Xuegong. Its screen wall, Lingxing Gate and east and west archway form a square. The pool in front of Lingxing Gate is named Panchi which is a unique composition of Confucius Temple. Visitors can enjoy the scene of Qinhuai River from the north bank of the Panchi.

The general design of Confucius Temple Qinhuai Scenic Area stresses traditional Chinese architectural elements and covers or forbids modern stuffs such as air-conditioner, and al-alloy doors and windows.

夫子庙建筑既有明清风格，又有庙、市、街合一的特色。由大成殿、学宫、江南贡院荟萃而成，是秦淮风光的精华。临河的贡院街一带则为古色古香的旅游文化商业街，同时按历史上形成的庙会的格局，复建了东市场、西市场，周围茶肆、酒楼、店铺等建筑也都改建成明清风格。

庙的位置或在学宫的前部，或偏于一侧。前设照壁、棂星门和东西牌坊形成庙前广场，棂星门前设以半圆形水池，称为"泮池"。泮池是大成殿的特有形制，源自《周礼》，而夫子庙凿秦淮河为泮池，是唯一利用河道作为泮池的遗例。岸北为石栏，有"天下文枢"牌坊，游人至此可凭栏小憩，一览秦淮河风光。

整体设计强调粉墙、黛瓦、花格窗、马头墙等建筑元素的融入，同时对现代建筑符号，比如空调外挂机、铝合金门窗等予以遮挡或取缔。修缮的建筑按照"修旧如旧"的原则，采用原有建筑材料和色彩。

Major Commercial Activities 主要商业业态

After years of repairing, various shops, book stores, teahouses and restaurants of local traditional features are built on both sides of Qinhuai River from Taoye Ferry in the east to Zhonghua Gate in the west.

Confucius Temple is the headstream of snacks in Nanjing and it has become the snack center of the city since the Six dynasties. From Ming and Qing dynasties to Republic of China, this area became more prosperous for the springing up of lantern ships. There were many famous shops like Kuiguang Ge, Jiangyou Ji, Xueyuan and so on. After the victory of Anti-Japanese War, Jiangyou Ji, Liufengju and Little Paris are called the Top Three in Confucius Temple.

Since the 1980s, Confucius Temple has become flourishing. From December 1987, Gourmet Month is held once in the city each year and later twice. During the Gourmet Month, top chefs from all over the country are invited here to demonstrate their talents.

The featured snacks in Confucius Temple are favored by visitors form all over world. 19 kinds of snacks acquired awards or honorary titles of provincial level or ministerial level.

经过多年的修复，在东起桃叶渡、西抵中华门、1.8千米的秦淮河两侧，兴建了高低错落、富有地方传统特色的河厅河房、歌楼舞榭以及商业街上众多的书肆、小吃店、茶馆与酒楼，并在河上恢复了绝迹多年的"秦淮画舫"。

夫子庙是南京小吃的发源地，早在六朝时期，就成了全市的小吃中心，明清至民国时期，随着秦淮灯船的兴起而更加繁华。小小的夫子庙地区，有大大小小20家小吃店，成了南京小吃集大成之地。名店有奎光阁、新奇芳阁、蒋有记、雪园、永和园、六凤居、五凤居、德顺居、龙门居等，风味独特的月来阁则位于泊在秦淮河上的一条画舫上。抗日战争胜利后，蒋有记、六凤居和小巴黎合称为夫子庙三家。蒋有记以善制牛肉锅贴名噪金陵；六凤居以豆腐脑、葱油饼著称；小巴黎以女侍招客，颇有洋风。

20世纪80年代以来，夫子庙小吃空前繁荣兴旺，昔日的老字号永和园、奇芳阁、蒋有记、奎光阁、六凤居、包顺兴（今瞻园面馆）等得到发扬光大。从1987年12月起，南京市每年举办一次"美食月"活动，后发展为一年举办两次食品节。食品节期间，实行开放式经营，邀请北京、天津、南京、成都、昆山、无锡、扬州、镇江等地的名师前来献艺。

夫子庙风味小吃深受海内外游客喜爱，有19个品种获得了省级和部级奖励或荣誉称号。前国家主席江泽民品尝夫子庙小吃后题词"十里秦淮千年流淌，六朝胜地今更辉煌"。

Food & Beverage 餐饮类

Jinlingchun

Located at No.79 Shiba Street, Jinlingchun is an old famous brand of over one hundred years history. It used to be one of the largest and first-degree restaurants. Its full name is "Jinlingchun Chinese-Western Restaurant".

In the Republic of China, Jinlingchun became a place for the citizens to serve significant guests. But during the Anti-Japanese War, it was closed down and eventually ruined by the war. In the 1990s, Qinhuai District made a great effort to recover the appearance and develop the cooking culture of Confucius Temple. In October 1997, Jinlingchun Restaurant was completed with a building area of 2,800 m². Now, it is an appointed restaurant for receiving important guests of province, city and district.

金陵春

　　坐落于夫子庙大石坝街79号的金陵春酒楼是一家拥有百年历史的老字号。后厅紧靠秦淮河，门面有9开间宽，是南京规模较大、较高级的一家酒菜馆，全称"金陵春中西餐馆"。

　　民国时期，"金陵春中西餐馆"逐渐成为当时人们招待重要宾客的场所。当时以胡长龄等一代厨师为主所制作的菜肴被称为"京苏大菜"，而年轻一代的厨师被后人称为"京苏帮"。随着抗日战争战火的蔓延，"金陵春中西餐馆"一度关闭，最后毁于战火。20世纪90年代，秦淮区大力恢复夫子庙的旧貌和发展夫子庙的饮食文化。1997年10月22日，"金陵春"酒楼在夫子庙广场东侧文德桥边正式落成，建筑面积达2 800平方米。如今的金陵春酒楼是省、市、区接待贵宾的定点餐厅，先后获得"省级巾帼文明示范岗""南京市青年文明称号""江苏餐饮名店""三信三优"等荣誉。

Juxianlou

Located at No.188 Gongyuan Street, Juxianlou has an elegant environment and a classic breath. Its building is grand and dignified. It can house more than two thousand people dining at the same time. There is a book bar on its second floor which is antique and unique. Its authentic duck blood vermicelli soup is both good looking and delicious, therefore it is very popular.

聚贤楼

聚贤楼位于夫子庙贡院街188号，环境雅致，且颇具古典气息。大殿体量宏大，造型轩昂，可同时容纳两千多人就餐。前廊步梁做工精细，位于餐厅二楼的书吧古色古香，别具一格。店内地道的鸭血粉丝汤，清亮的汤汁上洒上了"艳丽"的香油，并配有青绿的蒜花点缀，红、白、绿相间，赏心悦目，一天能卖出上百份。

Wanqinglou

It is divided into three parts. They are Hall of Nice Dishes, Hall of Local Delicacies, and Hall of Folk Customs. Its decoration is echoed with the Confucius Temple building group; its interior facilities are of first-class and modern breath.

晚晴楼

晚晴楼现有美食轩、风味轩、民俗轩三个部门，是专营晚晴"八绝"风味小吃的特色餐饮品牌。小楼青砖小瓦，粉墙坡屋，古朴典雅，与夫子庙古建筑群融为一体，内部设施一流，颇具现代气息。

Xianheng Hotel

It was a small hotel founded in 1894 by Zhou Zhongxian, Lu Xun's uncle. The hotel was facing the street with a square counter, several long tables and stools. It offered beans flavored with aniseed, peanuts and so on. Most of its customers were workers. Though the hotel was small, Zhou Zhongxian paid a great attention to its name and eventually picked a word from Yijing. Its unique operation form containing humanistic characteristics of Jiangnan has become the cultural base of Xianheng Hotel.

咸亨酒店

清光绪二十年（1894年），鲁迅先生的堂叔周仲翔等在绍兴城内的都昌坊口开设了一家坐南朝北的小酒馆。店铺临街，曲尺柜台，几张条桌、板凳，摆有茴香豆、煮花生之类的下酒小菜，顾客多为站着喝酒的"短衣帮"。酒馆虽小且简陋，但饱读史书的周仲翔对店名却十分讲究，几经斟酌，从《易经·坤卦》之"含弘光大，品物咸亨"句中，取"咸亨"两字为店名，寓意酒店兴隆，万事亨通，蕴含江南水乡、酒乡的人文特征和独特的经营风格，成为咸亨酒店的文化根基。

Metersbonwe

Shanghai Metersbonwe Clothe Co., Ltd. was founded in Wenzhou, Zhejiang in 1995. It engages in developing, design, producing and selling casual clothes targeting at young group of 16 to 25 years old. It proposes brand image of young energy and unique fashion.

美特斯·邦威

上海美特斯·邦威服饰股份有限公司于1995年创办于浙江省温州市。其主要研发、设计、生产、销售休闲类服饰，目标消费者是16～25岁的年轻人群，倡导青春活力和个性时尚的品牌形象。

Dacheng Hall 大成殿

It is the major hall of Confucius Temple hung with the largest portrait of Confucius in the country at the middle. The hall is furnished with fifteen ancient musical instruments used for worshipping Confucius, such as imitated chimes of 2,500 years ago. And there are archaic music performs on a regular basis.

大成殿是夫子庙的主殿，高16.22米，阔28.1米，深21.7米。殿内正中悬挂着一幅全国最大的孔子画像，高6.50米、宽3.15米；殿内陈设仿制2 500年前的编钟、编磬等15种古代祭孔乐器，定期进行古曲、雅乐演奏，使观众能听到春秋时代的钟鼓之乐、琴瑟之声，展现2 000多年前的古乐风貌。

Jiangnan Examination Hall 江南贡院

Built in 1168, it was specifically used for examination. Its site begins from Taoye Ferry in the east to Zhuangyuanlou in the west, and faces Qinhuai River in the south, and Jiankang Road in the north. Its area is about 300,000 square meters.

江南贡院始建于南宋乾道四年（1168年），初时，供县府学考试之用。明初定都南京，这里成为乡试与会试的考场。贡院东起桃叶渡，西至状元楼，南临秦淮河，北抵建康路，占地约300 000平方米。

Wuyi Lane 乌衣巷

Located in the southwest tens of meters away from Confucius Temple, it is a quiet and narrow lane. Residences of Wang Dao and Xie An, two famous ministers in Jin Dynasty, used to be located here.

乌衣巷位于夫子庙西南数十米，是一条幽静狭小的巷子，原为东晋名相王导、谢安的宅院所在地。旧时王谢子弟喜穿黑色衣服，因而得名。

Zunjing Pavilion 尊经阁

It is an ancient building which is dignified with double eaves, T-shaped ridge and gable roof. It used to be a classroom for giving lessons. It has collected the *Thirteen Confucian Classics* and the *Twenty-one History*.

尊经阁是一座重檐、丁字脊、歇山顶的三层古建筑，端正凝重、玲珑华丽，匾额由我国当代书坛女杰萧娴题写，意为"以经为尊"。古时候为讲堂，楼上藏有《十三经》《廿一史》等。

Xijin Ferry Street, Zhenjiang
镇江西津渡古街

街区背景与定位

Street Background & Market Positioning

History 历史承袭

The about 1,000 meters long Xijin Ferry Street was founded during Six dynasties. Xijin Ferry was called Suanshan Ferry during the Three Kingdoms, and Jinling Ferry in Tang Dynasty. It was originally adjacent to Changjiang River and gradually moved to the foot of Yushan Mountain, for the rise of river beach after Qing Dynasty.

Xijin Ferry has been a famous ferry of Changjiang River since the Three Kingdoms period. It is the only ferry linking Zhenjiang and the north of the river in Tang Dynasty. Therefore, it has a extremely significant strategic position.

Lu You, a famous poet in Song Dynasty had once stunned by thousands of sources of troops transported through Xijin Ferry everyday. Yu Shuzi, a poet in Qing Dynasty also wrote a poem to describe the bustling scene of Xijin Ferry.

　　西津渡古街全长约1 000米，始创于六朝时期，历经唐、宋、元、明、清五个朝代的建设，留下了如今的规模。西津渡，三国时叫"蒜山渡"，唐代曾名"金陵渡"，宋代以后才称为"西津渡"。这里原先紧临长江，清代以后，由于江滩淤涨，江岸逐渐北移，渡口遂下移到玉山脚下的超岸寺旁。

　　从三国时期开始，西津渡就是著名的长江渡口。镇江自唐代以来便是漕运重镇，交通咽喉。西津渡则是当时镇江通往江北的唯一渡口，具有极其重要的战略地位，自三国以来一直是兵家必争之地。

　　陆游途径西津渡时，曾对渡口每日运送上千的兵源感叹不已。清代诗人于树滋所写的诗更道出了西津渡渡口人来舟往的繁忙景观："粮艘次第出西津，一片旗帆照水滨。稳渡中流入瓜口，飞章驰驿奏枫宸。"

Location 区位特征

Xijin Ferry Street is located at Yuntai Mountain foot at the west of Zhenjiang. In ancient times, there was a Xiang Mountain at its east to ward off billowy sea tide, ancient Han Canal at its north and a precipice facing the river as a stable natural harbor. Now this area has most cultural relics and is the context of Zhenjiang.

　　西津渡古街位于镇江城西的云台山麓，是依附于破山栈道而建的一处历史遗迹。古时候，这里东面有象山为屏障，挡住汹涌的海潮，北面与古邗沟相对，临江断矶绝壁，是岸线稳定的天然港湾。至今，其仍然是镇江文物古迹保存最多、最集中、最完好的区域，亦为镇江历史文化名城的"文脉"所在。

Market Positioning 市场定位

Taking "restoring culture, inheriting history" as its developing concept, the market positioning of Xijin Ferry Street is to improve its tourism service function, realize its sustainable development and make it a beautiful city card of Zhenjiang and a famous tourism site at home and abroad.

　　规划以"复兴文化、传承历史"为开发理念，进一步完善街区旅游服务功能，实现街区的可持续发展和永续利用，使之成为镇江靓丽的城市名片、国内乃至国际上知名的旅游目的地。

Street Planning 街区规划

Taking "inheriting history, restoring traditional features, improving functions" as principle, the planning aims to carry out protection and construction projects — repairing historical sites such as Life-saving Society, Zhaoguang Stone Pagoda, Guanying Cave and so on, and building Suanshan Garden, Xijin Yayuan and so on.

This planning orderly shows the old town by gathering scattered cultural resources and presents recessive culture connotation to integrate traditional culture into modern life, in order to make the old town a flag and exhibition area of history and culture in Zhenjiang.

　　街区规划以"传承历史文脉、复兴街区风貌、提升街区功能"为原则，以"连接割断的文脉、保持古街的风貌、澄清误解的史迹、突出渡口的特色、打造整合的平台"为理念，并积极开展保护建设工程：修缮五十三坡道路西侧围墙、救生会、昭关石塔、观音洞、金陵渡小山楼等文物建筑；修建"蒜山游园""西津雅苑""长江渔港"等工程。

　　规划集中有序地展示了古城风貌，集聚了分散的文化资源，显现隐性的文化内涵，让传统的文化特色融入现代生活，使之成为镇江历史文化名城的标志区、文化特色的展示区、"本地人常来、外地人必到"的旅游区。

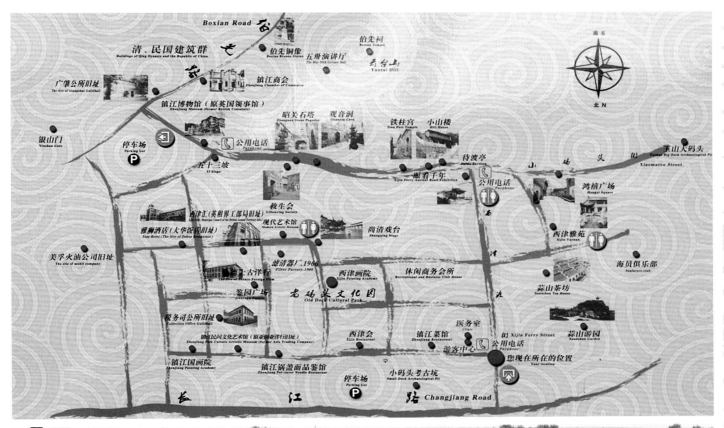

Street Design Features 街区设计特色

Zhenjiang is one of mountain forest cities which are rare in the country. The proper size and layout of Xijin Ferry Street fill itself with rich connotation and unique form. Though with the development of technology and society, the function of Xijin Ferry is gradually weakened, the general design of the street keeps it completely.

Buildings at the street are mostly relics of Ming and Qing dynasties. Window frames are all painted red and contrast to stone road in the street to form a distinct color effect. Buildings mostly apply wooden structure. Their lintels are carved with lane names such as "Chang'an Li", "Jirui Li" and "Dean Li". The houses are in ancient residential style with a courtyard in the middle.

Well-arranged two-floor buildings, overhanging eaves, distressed counters and other traditional decorations reveal the age of the old town. Walking in this antique and exquisite street is just like having a walk in a natural history museum.

镇江是国内为数不多的"山在城中、城在山中"的山林城市，而西津渡古街正是这一特点的集中体现，恰如其分的尺度和布局使其内涵丰富，形态独特。虽然科技的进步、社会的发展、环境的改变使西津渡逐渐淡化和削弱了作为渡口的功能，但街区的整体设计使它活化石般的风貌得以基本完整地保存了下来。

建筑多为明清时期的遗迹。飞檐雕花的窗栏一律漆成朱红色，给人以"飞阁流丹"之感，形成了"前店后寝"或"下面为店铺楼上为住家"的建筑形式，和街面的青石板路相互衬托，形成鲜明的色彩效果。建筑大多采用木结构，按里弄聚居。券门的门楣上镌刻着"长安里""吉瑞里""德安里"等字样。四合院中间为天井，反映出江南特有的"四水归堂"古民居建筑风格。

错落有致的两层小楼、翘阁飞檐、斑驳柜台、杉木十板门，娓娓诉说着"千年古渡、千年老街"的沧桑。"漫步在这条古朴典雅的古街道上，仿佛是在一座天然历史博物馆内散步。这里才是镇江旅游的真正金矿。"中国文物学会会长罗哲文先生更是把这里誉为"中国古渡博物馆"。

Major Commercial Activities 主要商业业态

In the street there are mainly restaurant, culture and leisure shops. There are also shops such as carpenter shops and mooring rope shops providing services for boatman.

街区商铺以餐饮、文化、休闲为主，兼有许多为船家服务的店铺，如木匠店、缆绳店等。

Featured Commercial Area 特色商贸区

Old Dock Cultural Park

Located in the northeast of Xijin Ferry Street, Old Dock Cultural Park is composed of ancient dock sites, concession building group remains and discarded manufacture factories. Dahua Restaurant, Shangqing Stage, Xijin Painting Academy and so on harmoniously coexist with traditional residences in the park.

Historical sites and remains such as Fan Fuxing Apartment, Yuelai Transport, factories for ships, Christianity Church and so on are repaired as old to keep the old appearance of the town, and therefore become a building group with unique features.

The business activities in Old Dock Cultural Park are antique yet fashionable. Restaurant, bar, teahouse, photographic studio, hotel and old-style private school here show diversified cultural charm.

老码头文化园

老码头文化园位于西津渡古街东北部，建筑构成有古代老码头建筑遗址、近代租界建筑群遗存和当代工业废弃厂房。在文化园内，税务司公馆、大华饭店、英租界工部局大楼、尚清大戏台、西津画院与传统民居和谐共生。

用于济渡行善的镇江义渡局旧址，用于消防的义渡洋龙局，为来往渡客服务的范复兴公寓、福春祥办馆、悦来搬运行、与造船有关的华美机器厂、邵记翻砂厂，从事建筑业的曹顺记营造厂，从事商业加工转口贸易的聚昌裕油厂，还有被毁的天主教堂等，经过"修旧如旧，以存其真"，成为古朴、富有特色的建筑群。

老码头文化园所集聚的商户业态中外兼具，古朴中透着时尚。饭店、酒吧、茶楼、影楼、旅馆、私塾等在粉墙黛瓦间彰显各种特色文化魅力。

Filter Factory 1966

Located beside Shangqing Stage, Filter Factory 1966 is a city leisure life house assembling leisure, entertainment, food and innovation. Its exterior keeps the original appearance of the old factory, while its interior is decorated fashionably and delicately to combine the antique with the modern perfectly. The factory is divided into several parts including a leisure food area, a business meeting area, an idea salon area, a fashion tea art area, an art exhibition area, audio visual area and so on.

滤清器厂1966

滤清器厂1966位于尚清戏台旁边，是集休闲、娱乐、简餐、创意展示为一体的城市休闲生活馆。外部保留了历史老厂房的风貌，内部进行了时尚、精致装修，让古朴与现代完美结合，包括休闲简餐区、商务会议区、创意沙龙区、时尚茶艺区、艺术展示区、视听体验区等。

商务休闲区设会议室、商务洽谈小包间和休闲区，轻松、简洁的空间设计让人的心情随之放松；视听区是一艘搁浅的木质帆船和旧机械组装的"变形金刚"，欧美怀旧电影海报、20世纪初美国老打字机将空间装饰得复古而精致，处处体现出一种休闲艺术的轻松之感。在一楼艺术长廊里，还可以欣赏各种艺术品、抽象派油画，同时为画展、摄影展等各类艺术展提供租赁空间。

Operation Measure 运营措施

By methods of financing, attracting investment, marketing sales and comprehensive operation of building archaistic residences and repairing important buildings, the project tries to realize the balance between investment and income of the project.

项目采取融资建设、招商引资、市场化销售的方式，将新建仿古民居、修缮重要建筑实施综合运作，滚动开发，力争实现街区投入与收益的平衡。

Food & Beverage 餐饮类

Vogue Coffee

Vogue Coffee has fashionable style and taste. Here visitors can taste authentic French goose liver, cocktail, coffee, pasta and so on.

Vogue Coffee

这里拥有时尚的格调和品位，可以品尝到正宗的法式鹅肝、鸡尾酒、咖啡、意大利面等。

Zhenjiang Restaurant

Located at Yuntai Mountain foot and right next to Xijin Ferry Street, the restaurant is a imitated Ming and Qing dynasties style building. The steamed bun with vegetable stuffing, made of first-class flour, is the most popular product and is sold in limited quantity.

镇江菜馆

镇江菜馆位于云台山下，紧挨着西津渡古街，是仿明清的古建筑，青砖墙面，有一种朴实温馨的感觉。该店的菜包子为一绝，每天限量销售，馅里加了白果仁，吃起来鲜香爽口，面皮用上等面粉制成，绵软而不粘牙。

Leisure 休闲类

Wenjin Club

By most strict technology and highest architectural standard, the club not only interprets the features of Zhenjiang, but also realizes the coexistence and co-prosperity of cultural relic protection and commercial interests.

问津会馆

问津会馆采用了最严格的技术水准和最高的建筑标准，在重新诠释镇江风貌的同时，实现了文物保护与商业利益共存共荣。

Chuanye Guild

It gathers accommodation, leisure and restaurant with comfortable and warm environment. It is a ideal place for holiday, tourism and party.

川页会所

川页会所集住宿、休闲、餐饮于一体，环境舒适、温馨，是度假、旅游、聚会的理想场所。

Accommodation 住宿类

Ferry Passenger Inn

Ferry Passenger Inn is a Jiangnan residence of Ming and Qing dynasties style made of brick and wood structure with 14 antique and exquisite rooms. Various art works are orderly and delicately furnished at every corner. There are open-air coffee seats, a Sunshine Book Bar, fashionable photo walls, antique mailbox and so on.

渡客客栈

渡客客栈是一栋砖木结构的明清式江南民居，共有典雅精致的客房14间。粗瓷碗、搪瓷缸、老煤油灯等各类艺术精品有序而别致地布满了各个角落，还配有露天咖啡座、阳光书吧、时尚照片墙、复古邮筒等。

Daidu Pavilion 待渡亭

It is a place where people welcome guests or families and see them off, take shelter from rain or wait for boat in ancient times.

待渡亭是古人迎来送往、小憩避雨、等待摆渡的场所，传说当年乾隆皇帝也曾经在这里停留。

Guanyin Cave 观音洞

It was dug on the massif beside Zhaoguan Pagoda to remember with gratitude the grace of Guanyin, the mercy Buddha and express people's good wishes.

观音洞是人们为了感念观世音菩萨的恩德，在昭关石塔旁的山体上开凿而成的，表达了人们祈祷平安的美好愿望。

Iron Post Temple 铁柱宫

It is for enshrining and worshipping the Xu Xun, the leader of Taoism in Jin Dynasty. Inside the temple are relics and remains of Taoism.

铁柱宫又称铁柱行宫，因供奉两晋道教明派许祖（许逊）真人而得名。洞内有石凿像台、供台平台、烧香池以及历代入洞台坡、道路遗迹，出土了若干具道教色彩的文物。

Shangqing Stage 尚清戏台

There are six upturned eaves of foreground and background which just like big wings skittering over water surface. There are twelve doors in front and five doors at sides which are covered with red paint and painted with patterns like mountains and characters.

尚清戏台前后台有6处翘檐，宛如大鹏的巨翅轻轻掠过水面。正面有12扇门，侧面有5扇门，都涂以鲜艳的红色油漆，上面绘有山水、人物等各种图案。

Small Dock Historic Site 小码头遗址

It is a hundred meters away from Yuntai Mountain in the north and is two to six meters under ground. Lots of porcelain and metal wares for daily life are unearthed here which reveal human activities during Qing Dynasty.

小码头遗址位于云台山北侧近百米处，曾被深埋在地下2～6米，坐南朝北。该地还曾出土了大量生活用瓷及铜、铁器物，展现了清代救生会、义渡局时期的人文活动，展示出富于特色的码头文化。

Zhaoguan Pagoda 昭关石塔

It is a stone pagoda over the street built in Yuan Dynasty. It is about five meters tall and is made of bluestone by parts.

昭关石塔是一座建造于元代的过街石塔。塔高约5米，分为塔座、塔身、塔颈、十三天、塔顶5部分，全部用青石分段雕成。

Zhenjiang Writer's Association Salon 镇江市作家协会沙龙

It is a professional group organized by literature worker in Zhenjiang voluntarily. It is also an intermediary for writers and literature workers to contact with society and market.

镇江市作家协会沙龙是镇江市文学工作者自愿结合的专业性人民团体，也是作家和文学工作者联系社会、面向市场的中介和依托，现有会员316人。

Zhufang Seal Society 朱方印社

It uses wood and computer, laser engraving to combine art of seal cutting with modern manufacture, therefore to realize minor appreciation moving toward democratic art.

朱方印社以木材作为印材，使用电脑、激光雕刻技术，使篆刻艺术与生产、生活相结合，走向市场，真正实现从"小众欣赏"走向"大众艺术"。

引导指示系统

Cangqiao Zhijie, Shaoxing
绍兴仓桥直街

History 历史承袭

Yuezicheng conservation district of historic sites is named after Guyuecheng. There is a Fushan Mountain, the biggest mountain in Shaoxing which is surrounded by an ancient moat, Huanshan River on the east and north sides, and the "Yuewang Stage" where Goujian — the king of Yue had meetings with ministers. Cangqiao Zhijie is located in this conservation district. It is like a masterpiece in this "museum without fence".

越子城历史文化保护区因位于古越城的原址而得名，保护区内有绍兴城内第一大山——府山，山的东侧与北侧环绕着旧时的护城河——环山河。在山拥水绕之中，越王勾践坐朝之处"越王台"矗立在街区中部。仓桥直街便处于越子城历史文化保护区范围内，如果说"流淌在2 500多年历史长河里的绍兴是一座'没有围墙的博物馆'"，那么，仓桥直街就是这个博物馆里的一件杰作。

Location 区位特征

Cangqiao Zhijie is located at the southeast foot of Wolong Mountain in Shaoxing. It includes several traditional lanes and a completed traditional residential area with mountain, water and the "Yuewang Stage". Embellished with historical sites, this street has mystery, profound and antique landscape features.

仓桥直街位于绍兴市卧龙山东南麓，街区中心线的环山河北起胜利西路，南至鲁迅西路。围绕山、水、台这三个核心景观的是几条传统的小弄及成片保护完整的传统民居区，具有人文背景的历史遗迹点缀其中，共同形成了街区神秘、幽远、古朴的景观特征。

Market Positioning 市场定位

Featured by the ancient city style, Cangqiao Zhijie with traditional residences as its main contents is a historic cultural street integrating residence, business and tourism.

仓桥直街是以古城风貌为特色，以传统民居为主要内涵，集居住、商业、旅游于一体的历史文化街道。

Street Planning 街区规划

General policy of the planning is to show the style of the ancient city, present its cultural connotation, improve residential environment, activate business tourism. The street is composed of river channel, folk residences and streets.

Taking "focal protection, reasonable reservation, partial modification and general improvement" as concept and following the principle of "restoring the old as old", the street keeps the original life scene.

项目规划总体方针：展示古城风貌，体现文化内涵，改善居住条件，激活商贸旅游。项目由河道、民居、街坊三部分组成，建筑面积53 892平方米，涉及总户数858户，其中有43个不同形式的台门。

以"重点保护、合理保留、局部改造、普遍改善"为理念，按照"修旧如旧、风貌协调"的原则修缮老街、老宅，保留了绍兴水乡民情生活的原生态；对古建筑等宝贵的历史文化遗产进行重点保护；对富有绍兴特色的建筑物进行合理保留；对具有较为典型的台门街坊逐步进行改造；对富有绍兴特色的街景、河道进行特色改造；对人民的生活水平进行普遍改善。

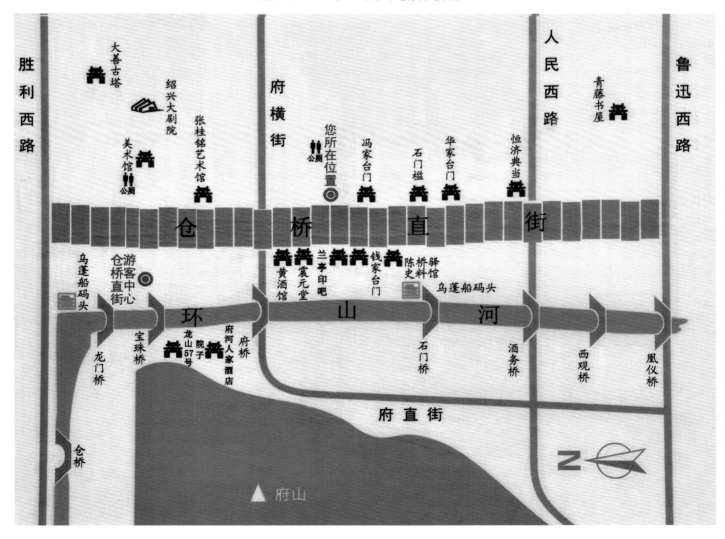

Street Design Features 街区设计特色

On both sides of river channel are mostly residences which are built during the late Qing Dynasty and early Republic of China, and reflect the traditional building features and living customs of Shaoxing. Traditional shops and restaurants at both sides of the street introduce local living customs to visitors as much as possible.

Buildings are mostly traditional residences of Shaoxing in the early Republic of China with gentle colors and comfortable sizes. They reflect the residential characteristics in certain natural and social conditions.

Corridors and quays are equipped along the river channel revealing that people depended on river channel in the past. Buildings away from the river are organized by courtyards of different sizes with changeable space layout.

Meanwhile antique slab bridges such as Cangqiao Bridge, Longmen Bridge, Shimen Bridge from south to north add charm to this street. General design of the street keeps its original appearance totally to not only present its reality of history and completion of style, but also continue its traditional life.

河道两旁以水乡民居为主，大多建于清末民初，其中有各式台门43个，集中反映了绍兴的传统建筑特色与民情习俗。街道两旁开设传统商店与餐馆，为让游人尽可能多地了解绍兴风情习俗，街道中还开设了越艺馆、黄酒馆、戏剧馆、书画馆等。

建筑为一河无街的格局形式，主要以民国初期绍兴水乡传统民居为主，色彩平和，空间尺度宜人，形成了较为完整统一的历史风貌，集中地反映了在特定的自然与社会条件下民居建筑的特色。

临水建筑设沿廊、埠头，以传统的绍兴木格子窗为主，沿河的阳台则恢复成披檐，反映了过去人民生活对水道的依赖。不临水的建筑由多户住宅形成连片建筑，以大小不等的院落（天井）组织平面，并有狭窄廊道相连，这样的街坊布局空间变化较多。

同时，自北而南依次架有仓桥、龙门桥、宝珠桥、府桥、石门桥、酒务桥、西观桥、凰仪桥等传统古老石板桥，增添水城氛围。整体设计原封不动地保持原貌，不仅体现了"历史真实性""风貌完整性"，更重要的是延续了传统生活的本真。

Major Commercial Activities 主要商业业态

Just like the springing up of tourism, the business in Cangqiao is also an unintentional outcome. People of Shaoxing have a tradition of doing business with a small capital at their forecourt in all ages. Therefore, the shops in this street are mainly small traditional featured shops. So far, there are more than 350 shops.

　　和旅游的兴起一样，仓桥的商贸亦为"无心插柳柳成荫"。绍兴人从古至今，就有在家的前院做小本生意的传统，因此，街区的商铺主要以传统特色小店为主。至今，整条街上已经有350多家商铺。

Food & Beverage 餐饮类

Yanyu Teahouse

Yanyu Teahouse is located in the southwest of Shaoxing City Plaza, and is founded by famous Shao opera artist Wang Zhenfang and his son. The teahouse displays folk classic art collections for looking back to Shaoxing tea culture. And it sells China's famous tea such as Longjing, Biluochun, Wulong and various flower tea. Famous tea made with water from Shunjiang River is pure, fresh and tasteful. The teahouse also provides tea ceremony show of profound cultural essence to perfectly combine Chinese tea culture with old Yue cultural art, therefore, creates a fragrant, quiet, interesting tea art environment.

雁雨茶艺馆

雁雨茶艺馆位于绍兴城市广场西南侧、卧龙山麓，由著名绍剧艺术家十三龄童及其子开办，馆内陈列着多年积累的民间古典艺术之精品，是追忆绍兴茶文化陈迹的好地方，也是素有集名城、名人、名艺、名茶之称的江南"品茶真迹"。馆内经营中国各大名茶："龙井"纤细俊秀，"碧螺春"柔曼娇嫩，"乌龙"苍老遒劲，"祁门红茶"汤似琥珀，众多花茶更是婀娜多姿，争相斗艳，清香扑鼻。名茶以会稽山源头舜江清泉泡沏，浓厚、醇和、鲜甜、韵味独特。这里还有蕴涵浓厚文化精粹的茶道表演，将中国茶文化和古越文化艺术完美合璧，构筑一隅清香、清静、清趣、清乐的茶艺境地。

Wulicai Restaurant

Wulicai restaurant originated from Xiangtan, Hunan. Its dishes are from dining-table of common families yet are luxury, for they are only made in rare important occasions. These dishes are famous for their real stuff, nutrition, health and pretty appearance.

屋里菜

屋里菜发源于湖南湘潭。"屋里"按湘潭方言的发音为"无哩",意为"家里的"。菜品来源于老百姓的餐桌,却是餐桌上的奢侈品,因其颇多讲究,一般家庭只有在少数很隆重的场合才能享用。菜品以真材实料、主料突出、讲求营养和健康、式样美观而出名。

Zhuangyuan Restaurant

Zhuangyuan Restaurant was founded in 1785 and was famous for its authentic Ningbo cuisine. The exterior of its building is antique, while its interior environment is more fashionable with floral wall paper, soft sofa, long table which are in exquisite layout.

状元楼

状元楼创建于清乾隆五十年(1785年),以经营正宗甬帮菜扬名。相传,曾有两位举人赴京赶考路经甬城,上这家酒楼临窗对江小酌,酒酣耳热之际,跑堂送上一盘"冰糖甲鱼"。两人看盘中青黄相映,油汁紧裹鱼块,入口绵糯,香、甜、酸、咸各味俱全,禁不住赞不绝口。与外建筑古色古香不同的是内部环境显露些许时尚,有小碎花墙纸、软沙发、长桌子,布局雅致。

Narcissus Restaurant

"Narcissus", which means original ecology, is always a compliment for aquatic products. Narcissus Restaurant is a propitious place for dates, parties and business banquets.

水仙酒楼

　　"水仙"是对水中鲜活产品的美好赞誉，意味着原生态，意味着水中仙品。水仙酒楼是情侣约会、家庭聚餐、朋友相聚、商务宴请的吉祥之地。

Bus Stop

Customers can read, taste coffee here. Its furnishings are very unique with two floors and a balcony beside the river.

Bus Stop

　　在店内可以看书、喝咖啡。这里布置很特别，上下两层，后面有阳台靠着河边。

Health Care 医疗类

Zhenyuantang

Zhenyuantang was founded in 1752. With more than 260 years history, it always does great in its business. It is said that the founder of Zhenyuantang made his fortune by a stall at the beginning and because of his good reputation built a solid family estate of Zhenyuantang.

震元堂

　　震元堂初创于清乾隆十七年（1752年），是绍兴诸店之魁，迄今已有260多年的历史，其营业历来鼎盛，久而不衰，被誉为"店运昌隆三百载，誉满江南数一家"。相传创始人杜景湘原是摆药摊起家，因经营有方，信誉卓然，遂得以创下震元堂坚实的基业。

2003 – UNESCO Asia-Pacific Heritage Awards

The project gives priority to repairing the street and respects its origin rather than pull it down and rebuilt it. The protection idea and concept totally agree with heritage protection principle.

2003年，联合国教科文组织亚太地区遗产保护奖

　　理由：绍兴仓桥直街的保护思路，非常符合遗产保护方法，即"修缮为主，尊重原真性，而不是推翻重建"。

绍兴仓桥直街荣誉

Honors

Gulangyu Island, Xiamen
厦门鼓浪屿

街区背景与定位

Street Background & Market Positioning

History 历史承袭

Gulangyu Island was named after a stone called "Gulang Stone" on the island which made the sound when billow hit it.

Gulangyu Island used to be sparsely populated for a long time in history. In the late Ming Dynasty national hero Zheng Chenggong once stationed troops here. After Opium War, thirteen countries including Britain, America, France, Japan, Germany and so on established consulates on the island. Meanwhile, merchants and missionaries came to the island to build residences, churches, foreign firms, hospitals, schools and so on. Some overseas Chinese built villas here in succession. Gulangyu Island ended its colonial history of over one hundred years after the victory of Anti-Japanese War.

　　鼓浪屿原名"圆沙洲"，别名"圆洲仔"，明代时改称"鼓浪屿"。因岛西南方海滩上有一块两米多高、中有洞穴的礁石，每当涨潮水涌，浪击礁石，声似摇鼓，人们称"鼓浪石"，鼓浪屿因此而得名。

　　在历史上，鼓浪屿曾长期是一个人烟稀少的荒岛。岛上多为半渔半农经济，最初的房屋也多是十分简陋的民房。明末，民族英雄郑成功曾屯兵于此。鸦片战争后，英国、美国、法国、日本、德国、西班牙、葡萄牙等13个国家曾在岛上设立领事馆，同时，商人、传教士纷纷踏上鼓浪屿，建公馆、设教堂、办洋行、建医院、办学校，成立"领事团"，设"工部局""会审公堂"，把鼓浪屿变为"公共租界"，一些华侨富商也相继来此兴建住宅、别墅。抗日战争胜利后，鼓浪屿才结束了一百多年殖民统治的历史。

Location 区位特征

Gulangyu Island located in the southwest of Xiamen Island is the largest satellite island of Xiamen. Because of hits of billow all the year round, there are many deep and secluded valleys and cliffs; houses stand side by side among the tropical or subtropical woods. With pleasant climate and without noise of vehicles, this island is praised as "a garden on the sea". For the history reason, it is also called "a display of international buildings".

　　鼓浪屿位于厦门岛西南隅，与厦门市隔海相望，是厦门最大的一个卫星岛。此地岩石峥嵘，挺拔俊秀，因长年受海浪扑打，形成许多幽谷和峭崖，沙滩、礁石、峭壁、岩峰，相映成趣；楼房鳞次栉比，掩映在热带、亚热带林木里，日光岩奇峰凸起组成了一幅美丽的画卷。岛上气候宜人，四季如春，无车马喧嚣，素有"海上花园"之美誉。由于历史原因，中外风格各异的建筑被完好地汇集、保留，故又有"万国建筑博览"之称。

Market Positioning 市场定位

By integrating history, culture and nature, a commercial service system of Gulangyu's features with reasonable layout and completed functions is formed to create a comprehensive island cultural tourist area gathering sightseeing, vacation, travel, shopping and entertainment.

　　项目规划融历史、人文、自然景观于一体，形成既满足外来游客需求、又方便岛上居民生活使用，布局合理、功能完善、具有鼓浪屿特色的商业服务体系，打造集观光、度假、旅游、购物、休闲、娱乐于一体的综合性海岛风景文化旅游区。

Street Planning 街区规划

1)Strictly control the business: the island is classified into three types, commercial concentration area, controlled guidance areas and strictly controlled area. Among them, in strictly controlled area are mainly scenic zones.

2)Conduct commercial layout: there is an east-west commercial belt and a west-north seaside featured commercial belt.

3)Centralize restaurants: to reorganize the existing snack street to expand its area and to build a food concentrated area at Fuzhou Road.

4)Remold the farmers market: Longtou Farmers' Market will be remolded to a fresh food supermarket; Neicuo Community will be also equipped with a community commercial center.

5)Strengthen the examine and approve standard: to control and avoid disorder business which disagrees with the aim of application for the list of world heritage.

　　1)严格控制商业：全岛将被划分为商业集中设置区、控制引导区和严格控制区三类，严格控制区主要为风景区、申遗核心要素保护区、风貌建筑等集中的区域，区内严格控制商业设施的进入。

　　2)引导商业布局：向西通过泉州路—笔山洞—内厝澳路—内厝码头形成东西向商业带；向北通过福州路—延平路延伸至三丘田码头，形成西北向滨海特色商业带。

　　3)集中餐饮街市：整顿福州路现有小吃街，扩大道路面积；建设福州路餐饮集中区，龙头路定位为精品商业购物街，其餐饮业态将尽量集中到福州路餐饮集中区内。

　　4)改造农贸市场：龙头农贸市场将改造成生鲜超市，内厝社区也将配置社区商业中心一处，以现有集贸市场为中心进行改造提升。

　　5)前移审批关口：启动由鼓浪屿管委会牵头，工商、食品卫生、公安、消防等部门共同参与、统一受理的联审制度。联审制度将审批"关口"前移，从源头上控制与申遗目标不符的商业无序乱象。

Tourism Components Layout 旅游要素布局

In the planning, "food, accommodation, route, shopping and entertainment", these five elements are systematically analyzed and their layout is adjusted.

Food — to reorganize the existing snack street on Fuzhou Road and expand its area to build a west concentrated open-air food & beverage area; according to Zhaohe Mountain to rebuild the northeast to an open-air music bar.

Accommodation — to position Lujiao Road, Fuxing Road and so on as high-end area and remold featured buildings in these streets to family inn.

Route — to the traffic on the island is mainly on foot. Its route is in ring radiate system to link main scenic spots. Some key nodes are equipped with commercial facilities.

Shopping — to set up Longtou Road shopping area, Neicuo shopping street and Sanqiutian Ferry shopping spot. Some of old famous shops are restored to form a Gulangyu unique commercial atmosphere.

Entertainment — there are culture exhibition, art performance and sport complex.

在规划过程中，对"吃、住、行、购、娱"等五大旅游要素进行了系统分析，并对布局进行调整。

吃——整顿福州路现有小吃街，扩大道路面积，结合小浪荡山现有建筑及场地条件，建设西部露天餐饮集中区；结合兆和山东北侧现有条件，将其改造为独立的露天音乐酒吧场所。

住——东部沿鹿礁路、复兴路等街区的家庭旅馆定位为高端区，将风貌建筑改造为家庭旅馆，并严格注意开发与保护的关系。

行——内部交通主要以步行为主，步行结构为环状放射结构系统，串联各主要景区；在关键节点设置部分商业设施。

购——设置龙头路旅游购物街区、内厝旅游购物街、三丘田码头购物点；恢复部分鼓浪屿历史上的老字号商店，以形成具有鼓浪屿特色的商业氛围。

娱——文化展览场所、艺术表演场所、体育运动场所三者兼具。

Street Design Features 街区设计特色

The street is a combination of "points, lines and surface" and featured business.

"Points" include two communities that are community commercial centers remolded from Longtou Farmers' Market and Neicuo composite market; two visitor centers; two tourist commercial nodes at Sanqiutian and Neicuo Ferry; and commercial service facilities set according to scenic spots layout.

"Lines" include linear commercial streets such as Quanzhou Road, Zhonghua Road, Neicuoao Road and Fujian Road.

"Surface" refers to Longtou Road Commercial Street.

"Featured business" presenting featured business spaces of nature, history, music and culture includes seaside leisure restaurant zone, featured style building zone, family inn zone and music culture exhibition zone.

Buildings here generally adopt local stones and red bricks. Their materials are selected and their shapes are unique. The facades of buildings are decorated with Romanesque columns and slope roofs of totally different shapes. Some of the buildings imitate ancient palatial architecture while blending European architectural technique to create a compromise style of unique appearance.

街区整体形成"点、线、面"及特色商业相结合的布局。

"点"包括两个社区中心，即龙头农贸市场、内厝综合市场改造为社区商业中心；两个游客服务中心，除现有龙头路游客服务中心之外，新增内厝旅游码头游客服务中心；在三丘田和内厝码头各设置一处旅游商业节点；结合景区景点分布，设置出景区（点）商业服务设施。

"线"包括泉州路、中华路、内厝澳路、福建路等线形商业街。

"面"指龙头路商业街区。

"特色商业"为体现鼓浪屿阳光、沙滩、海岸、历史风貌、音乐、人文等的特色商业空间，包括滨海休闲餐饮空间、特色风貌建筑空间、家庭旅馆空间、音乐文化展示空间等。

建筑一般选用当地的石材或红砖，用料考究、造型别致，且多采用圆拱回廊、清水红砖、红瓦坡屋面，并用柚木楼板、花砖铺地，辅以奇异别致的琉璃瓶花格。建筑的各个立面常精雕细刻罗马式大型圆柱和结构造型迥然不同的多坡屋顶。有的建筑仿照古代宫殿式建筑并融合西欧建筑造型手法，形成外形独特、屋檐线条奇异的折中式建筑风格。

Major Commercial Activities 主要商业业态

Food & beverage: featured snacks, Chinese and western food, food stall, seafood restaurant and so on.

Shopping: Longtou Road shopping area mainly provides local specialties, culture boutiques, creative products; Neicuo shopping street and Sanqitian Ferry shopping spot provide local specialties and tourist souvenir; scenic spots mainly offer souvenir, beverage and retails excluding local specialties.

Culture exhibition:book store, library, and museums.

Art performance: Haitiantanggou Mansion, music hall and seaside leisure music bar.

Sport complex: Dadeji Seaside Resort and Gangzihou Seaside Resort.

餐饮：包括特色小吃、中西餐点、大排档、海鲜酒楼等。
购物：龙头路旅游购物街区内以土特产、文化精品、创意产品为主；内厝旅游购物街及三丘田码头购物点以销售土特产、旅游纪念品为主；旅游景区、景点主要以纪念品、饮料、零售为主，不设置土特产销售点。
文化展览场所：包括书店、图书馆、乐器博物馆、风琴博物馆、钢琴博物馆、郑成功纪念馆、近代历史博物馆、内厝澳社区博物馆等。
艺术表演场所：包括海天堂构、音乐厅、滨海休闲音乐酒吧。
体育运动场所：包括大德记海滨浴场、港仔后沙滩浴场等。

Food & Beverage 餐饮类

Pan Xiaolian

Located at No.8 Longtou Road,Pan Xiaolian's environment is elegant with Southeast Asia style. It is a characteristic shop featuring plain yogurt.

潘小莲

潘小莲位于鼓浪屿龙头路8号，环境优雅，具有东南亚风情，是一家主营原味酸奶的特色小店。

Huangshengji

Huangshengji is one of the most famous old brands in Xiamen. It is 166 years old. For good material and skillful craftsmanship, its dried meat floss is good-looking and unparalleled delicious. It has several branches. This brand became very famous in the Southeast Asia.

黄胜记

"黄胜记"是厦门最具知名度的中华老字号之一，至今已有166年的历史。因选料精良，配料独特，加之精工细作，所产肉松色泽金黄，灿若黄金，其味香鲜，无与伦比，并创立了"黄金香送记""黄金香佑记""黄金香胜记"等分号。黄金香在东南亚地区享有盛誉。

Huafeng Hotel

This building has a strong European style. Arch, balcony window are extraordinary splendor and full of classicism and romanticism.

华风宾馆

华风宾馆建筑具有强烈的欧陆风格，门楼壁炉、阳台、勾栏、突拱窗等都别具特色，洋溢着古典主义和浪漫主义的色彩。

Consulate Inn

It is located at Lujiaoding facing Gulangyu Piano Ferry. It is a European-style villa built by the second consul of Britain in Xiamen in November 1844. The third floor is a holiday inn of Chinese-Western style. The beautiful scene of Lujiang River outside the window, and the culture breath and fragrant coffee inside the window provide visitors with unexpected taste.

卡斯特旅馆

卡斯特旅馆坐落于鼓浪屿钢琴码头对面的鹿礁顶，1844年11月，英国驻厦门第二任领事建立了这座独立的欧式别墅；1998年外交部根据原来的建筑风貌进行整修翻新。三楼为中西合璧风格的度假旅馆，提供特色古典房4间、浪漫房3间、欧式豪华房3间。古式木窗外迷人的鹭江美景、窗内的复古文化气息和香浓咖啡，给旅客以意想不到的体验。

Silly Girl Coffee Hotel

Silly Girl Coffee Hotel used to be London Missionary Society, and is a very beautiful British - style building. It has totally ten rooms, a 200 m² yard and a ring-shaped British ambulatory with public supports of audio-video rooms, studies, cafeteria, self-service laundry and so on.

喜林阁咖啡旅馆

喜林阁咖啡旅馆前身为伦敦差会姑娘楼，建于1842年，是非常漂亮的英式建筑。英式回廊颇具异国风情。旅馆共有10间客房，同时有200平方米的院子和一个环形的超大英式回廊，提供影音室、书房、自助餐厅、自助洗衣房等公共配套。

Leisure 休闲类

Sanyou Holiday Tourist Town

It is the largest retail center with most completed facilities and most advanced management in Gulangyu Island. It gathers tourism, leisure, food, shopping and entertainment. Its main building integrates European style with arcade buildings of Fujian and constructs a classic one. It aims to create a unique leisure culture and comfortable tourist shopping environment for tourists.

三友假日旅游城

三友假日旅游城是鼓浪屿目前最大、设施最完备、管理最先进的商业零售中心，集旅游、休闲、美食、购物、娱乐于一体。主体建筑把欧陆风格和闽南独有的骑楼相融合，构成了中西合璧的经典作品。全开放式设计的卖场亦极具特色，以"天天欢乐、分享快乐"为经营理念，着力为游客营造独特的节庆休闲文化和舒适的旅游购物环境。

Tianxia Silk

It mainly sells silk products including national clothes, cheongsams, scarves and so on. Their commodities with top-class quality of various types are mainly from Hangzhou and they might be another wonderful experience apart from food in Gulangyu Island.

丝绸天下

丝绸天下主要销售丝绸产品，包括具有民族风的服装、旗袍、丝巾等。产品主要来源于丝绸之乡杭州，质量上乘，包括富丽堂皇、轻薄飘逸、柔软滑爽等各种类型，使其成为继鼓浪屿美食之后又一不容错过的体验。

Cultural Facilities 文化设施

Union Church 协和礼拜堂

In the middle of 19th century, many westerners came to Gulangyu Island for job and study. In order to hold Sunday worship, Christians donated to build the church at Lujiaoding which is the earliest church in Gulangyu Island and named "International Church". In 1911, the church was renovated and renamed as "Union Church".

19世纪中期，不少西方人来到鼓浪屿工作、生活。为了举行主日崇拜，各差会(美国归正教会、英国伦敦公会、英国长老会)的信徒纷纷捐款，在鼓浪屿的鹿礁顶（亦称上鹿礁）建造教堂。1863年，这座新古典主义风格的教堂竣工，时称"国际礼拜堂"，是鼓浪屿最早的教堂。1911年，教堂被翻建，改称"协和礼拜堂"。

Oriental Fish Bone Gallery 东方鱼骨艺术馆

It is the first professional gallery of fish bone art mainly exhibiting pictures composed of fish bone. Fish bone can be used for painting after 12 processes including cleaning, whitening, molding, curing, drying and so on. For its different shapes and varieties, fish bone can be transformed into not only vivid realistic painting, but also abstract painting which can express the meaningful artistic imagination through artificially conceptive arrangement.

东方鱼骨艺术馆是目前国内乃至国际首家专业鱼骨艺术馆，主要展示各类以鱼骨为材料所构思创作的画作。鱼骨材料一般经过脱肉、净白、去腥、去味、防霉、烘干、美味等12道工序，方可作画。由于鱼骨材料造型各异、变化万千，经过巧妙的构思和精心的搭配，既可以制作出栩栩如生的写实画，亦可以创作出抽象写意画。

Catholic Church 天主堂

Located at No.15 Cian Road and built in 1860, the Catholic Church can accommodate more than 500 people. It is a Romanesque church with double bell towers and credence tables at the middle and both sides. The middle credence is worshipping the Chinese Rose Madonna.

天主堂位于厦门市区磁安路15号，占地面积413平方米，1860年建成，可容纳500多人，为罗马式双钟楼教堂，正中及两边各有一个祭台，正中祭台供奉中华玫瑰圣母像，故该堂又称"玫瑰圣母堂"。

Zhongde Temple 种德宫

Zhongde Temple is a temple with three doors and two halls. Its site area is about 600 m² and building area is about 78 m². Under an old banyan tree, its environment is quiet and peaceful.

种德宫是一座"三门二进"殿宇，总占地面积约600平方米，建筑面积约78平方米，宫后及宫右另筑护厝，整座宫宇掩映于古榕树下，环境幽静。

鼓浪屿荣誉

Honors

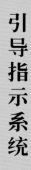